Lecture Notes in Control and Information Sciences

Edited by M. Thoma

For information about Vols. 1–42 please contact your bookseller or Springer-Verlag.

Lecture Notes in Control and Information Sciences

Edited by M. Thoma and A. Wyner

106

R. R. Mohler (Editor)

Nonlinear Time Series and Signal Processing

Springer-Verlag Berlin Heidelberg GmbH

ISBN 978-3-540-18861-2   ISBN 978-3-540-38837-1 (eBook)
DOI 10.1007/978-3-540-38837-1

2161/3020-543210

PREFACE

This volume provides a sample of relevant new results on dynamical nonlinear
statistical modeling and estimation which forms a basis for more effective signal
processing, decision, and control. While the research literature is rich in linear
Gaussian methodologies, new contributions to the most relevant area of nonlinear
and non-Gaussian processes have been scarce. Among the significant areas of
application for which such methodologies are needed are: economics, biology,
immunology, underwater acoustics, electric power generation, chemical process
control, and variable structure systems in general. The latter include adaptive,
intelligent, and decomposing mathematical structures or processes.

The present volume includes ten research papers on theory, computational
methods, and applications. The first two papers study filtering (i.e., state
estimation or model-based signal processing). In the first paper, inertial
navigation modeling and filtering are studied by Balakrishnan. Here, a linear
approximation is used to derive the Kalman filter. The second paper, by Kolodziej
and Pacut, introduces a new methodology for the detection of system structural
change such as may arise due to component failure, addition, or deletion of
processes (such as measurements in an array of receivers for various incoming
signals) or infection of an organ. Then, the nonlinear filter is based on the
latest model structure and nonlinear measurements. Spectral and bispectral methods
of analysis such as are appropriate for bilinear systems, BLS, are presented by
Rao. Spectral estimates for random magnetic reversals are studied here. From this
paper, BLS form a commonality with the second paper, if the nonlinear measurement
feedback synthesizes a control or decision function, and with subsequent papers.
Parameter estimation for bilinear time-series models and affine bilinear models are
studied next by Tang and Mohler and by Kumar. Recursive least-squares residual
comparisons and direct moment estimator methods are utilized to identify model
parameters. Applications which are presented here include appropriately maneuvering point object and stock price data analysis with reference also made to unemployment data - all for which linear models were less accurate forecasters.

In the study of population biology, Tong considers a changing structure which
arises physiologically. Model threshold evolves from food limitation and delay

time from development time. Such <u>threshold models</u> are further studied by Li, who utilizes the Akaike information criterion to estimate delay parameters. Variable-structure processes sometimes become <u>catastrophic</u> and the nonlinear series analysis of such catastrophes is analyzed by Cobb and Zacks in the next paper.

A new signal-processing methodology, which is based on first and higher-order convolutions, is presented by Bugnon and Mohler. The method leads to a generalized (higher-order) <u>eigenstructure generator</u> which may be convenient for certain <u>array processing</u> involving coherent and/or non-Gaussian signals. Finally, a basis is established by Farooqi and Mohler for further research on the application of bilinear time-series analysis to <u>experimental immunology</u>. Here, non-Gaussian experimental data of lymphocyte circulation leads to a stochastic-parameter dynamic model. Unfortunately, the conservation of cells for the recirculatory process leads to a lack of second-order (and higher) stationarity which, in conjunction with the vastly scattered minimum data set, makes model identification difficult.

It is impossible to acknowledge or to reference all of the important contributions to these topics by other researchers. Many such references are made in the individual papers in this volume. A few key related contributors include: Beneviste, Basseville, Nikiforov and Willsky (on change detection); Brillinger, Cartwright, Feigin and Tweedie, Gabr, Granger and Anderson, Hinich, Nichols and Quinn, Priestley, Rosenblatt (on non-Gaussian/nonlinear time series and spectral estimation). Among those recent ones not appearing in the references of this volume are Nikias and Raghuver ("Bispectrum Estimation," Proc. IEEE 75, 869-891, 1987) and Stensholt and Tjostheim ("Multiple Bilinear Time Series Models," J. Time Series Anal. 8, 221-233, 1987).

Ron Mohler
Corvallis, Oregon
November 1987

CONTENTS

ON THE APPLICATION OF KALMAN FILTERING TO CORRECT ERRORS DUE TO VERTICAL DEFLECTIONS
IN INERTIAL NAVIGATION

A.V. Balakrishnan
Electrical Engineering Department
UCLA
Los Angeles, CA 90024 U.S.A.

1. INTRODUCTION

There is a large literature on the application of Kalman Filtering to Inertial
Navigation Systems. (See e.g., [2], [7], [8], [12]). In this paper in order to
present a concrete problem, we shall deal only with Shipboard Inertial Navigation
Systems (SINS), and in particular, we shall discuss only the application of Kalman
filtering for correcting errors due to vertical deflection (VD) of gravity. After
briefly reviewing the basic physical theory, we shall formulate the main problems
and indicate some current engineering solution, emphasizing the potential unsolved
problems and research areas. Of particular mathematical interest here is the way in
which random fields enter a "time-domain" filtering problem.

2. BASIC PHYSICAL THEORY

We shall begin with a simplified analysis of SINS errors due to vertical
deflections of gravity, following [13]. The interested reader should consult the
references for more details. Thus let us consider motion confined to the meridian
plane. Let ϕ denote the true latitude so that the true acceleration a is given by:

$$a = R \frac{d^2\phi}{dt^2}$$

where R is the radius of the earth. Let a' be the acceleration indicated by the
shipboard accelerometer, so that

$$a' = R \frac{d^2\phi'}{dt^2}$$

where ϕ' is the indicated latitude. Let

$$\beta = \phi' - \phi$$

denote the error. Then with reference to Figure 1 we can readily see that

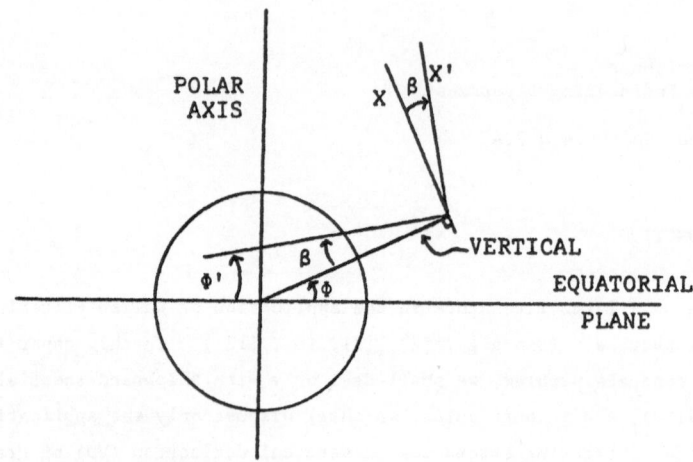

Figure 1

$$a' = \left(a + \frac{\partial w}{\partial (R\phi)} \right) \cos \beta + \left(g - \frac{v^2}{R} \right) \sin \beta + N(t)$$

where

w	—	the gravity potential
g	—	"gravity field" $\approx |\nabla w|$
v	—	$R\dot{\phi}$ — ship velocity
N(t)	—	accelerometer noise.

Now

$$v^2 \ll R$$

so that

$$g - \frac{v^2}{R} \approx g.$$

Further, β is small enough that we may take:

$$\sin \beta = \beta ; \quad \cos \beta = 1.$$

Further

$$\frac{\partial w}{\partial (R\phi)} = g\xi$$

where (by definition) ξ is the "north" component of the vertical deflection due to gravity.

Thus we have for the dynamics of the error angle β:

$$\frac{d^2\beta}{dt^2} = \frac{g}{R} \beta + \frac{g\xi}{R} + \frac{N(t)}{R} \ . \tag{2.1}$$

Note that

$$\xi(t) = \xi(\phi(t), \lambda(t))$$

where $\lambda(t)$ is the longitude, and

$$\xi(\phi, \lambda)$$

is a two-parameter field, $-\frac{\pi}{2} \le \phi \le \frac{\pi}{2}$, $0 \le \lambda \le 2\pi$, over the earth's surface.

Equation (2.1) describes the SINS error model -- North Channel. The complete error-model is obtained by including the "East Channel" and may be described by the "state-space" equation:

$$\dot{x}(t) = Ax(t) + N(t) + \zeta(t) \tag{2.2}$$

where

$$x(t) = \begin{bmatrix} p^N \\ \dot{p}^N \\ p^E \\ \dot{p}^E \end{bmatrix}$$

p^N is the North position (error)

$$p^N = R(\Delta\phi)$$

$$p^E = (R \cos \phi)(\Delta\lambda) \text{ (East position error)}$$

and

$$A = \begin{bmatrix} 0 & 1 & 0 & 0 \\ \frac{-g}{R} & -c_1 & 0 & (-2\Omega \sin L) \\ 0 & 0 & 0 & 1 \\ 0 & (2\Omega \sin L) & \frac{-g}{R} & -c_1 \end{bmatrix}$$

where, to be slightly more realistic, we have added some damping (c_1) and taken account of the Coriolis force ($2\Omega \sin L$).

$$N(t) = \begin{bmatrix} 0 \\ N_1(t) \\ 0 \\ N_2(t) \end{bmatrix}$$

where $N_1(\cdot)$, $N_2(\cdot)$ represent accelerometer "noise," usually assumed white. The most important term (for us) is the V.D. term:

$$\zeta(t) = \begin{bmatrix} 0 \\ g\xi \\ 0 \\ g\eta \end{bmatrix}$$

$$\xi(t) = \xi(\phi(t),\lambda(t))$$

$$\eta(t) = \eta(\phi(t),\lambda(t))$$

$$\xi(\phi,\lambda) = \frac{\partial w}{g\partial(R\phi)} = \text{North component of V.D.}$$

$$\eta(\phi,\lambda) = \frac{\partial w}{g\partial((R \cos \phi)\lambda)} = \text{East component of V.D.}$$

Let

$$\left.\begin{array}{l} \phi' = \phi + \Delta\phi \\ \lambda' = \lambda + \Delta\lambda \end{array}\right\} \qquad (2.3)$$

so that (ϕ', λ') is the SINS "output."

3. AN ENGINEERING SOLUTION

Before we consider the general case, let us look at an engineering solution to the "filtering" problem (of correcting errors due to V.D.) when an external aid is available. Thus suppose, in addition to (2.2)-(2.3), we are given:

$$\phi_e(t) = \phi(t) + N_3(t)$$

$$\lambda_e(t) = \lambda(t) + N_4(t)$$

where $N_3(\cdot)$, $N_4(\cdot)$ are errors modeled by white noise. Let

$$\left.\begin{aligned}
y_\phi(t) &= \phi'(t) - \phi_e(t) = \Delta\phi(t) - N_3(t) \\
y_\lambda(t) &= \lambda'(t) - \lambda_e(t) = \Delta\lambda(t) - N4(t) .
\end{aligned}\right\} \tag{2.4}$$

The problem then is to estimate $\phi(t)$, $\lambda(t)$ from (2.2), (2.3), and (2.4).

To proceed further with the error analysis, we need to know more about the two parameter fields $\xi(\phi,\lambda)$, $\eta(\phi,\lambda)$. The first question is: In what sense can we take them (and geodetic data generally) to be "random" since there is only one earth? The answer is that we are interested only in "local" (as opposed to "global") properties and the randomness is understood in the "local" sense -- for "small" (spatial) wavelengths. The physical area of interest is small enough that we may in fact use the "flat earth" theory:

$$R\Delta\phi \qquad \sim \Delta y \quad \text{(North)}$$
$$(R \cos \phi)\Delta\lambda \sim \Delta x \quad \text{(East)}$$
$$\text{Large R .}$$

Thus $\xi(\phi,\lambda)$, $\eta(\phi,\lambda)$ may be taken to be stationary random fields. Further simplification is afforded by taking

$$\mathring{\phi}(t) \sim v_1$$
$$\mathring{\lambda}(t) \sim v_2$$

where v_1, v_2 are constant, so that, as a consequence, $\xi(t)$, $\eta(t)$ are time-stationary processes. The spatial correlation function:

$$E[\xi(\phi+\Delta\phi, \ \lambda+\Delta\lambda) \ \xi(\phi,\lambda)] = R_\xi(\Delta\phi,\Delta\lambda)$$

being known (or assumed known from geodetic measurements), the correlation function of $\xi(t)$ is calculable, and that of $\eta(t)$, similarly. We assume that $\xi(\cdot)$, $\eta(\cdot)$ are Gaussian as well, with zero mean. Assuming furthermore that the spectral density matrix is rational, we can readily develop an on-line Kalman filter for $\Delta\phi$, $\Delta\lambda$, based on $y_\phi(t)$, $y_\lambda(t)$, and hence estimate $\phi(t)$ and $\lambda(t)$. A common choice for the correlation function of $\xi(t)$, $\eta(t)$ is

$$E[\xi(t+\Delta)\xi(t)] = \sigma_\xi^2 \, e^{-k|\Delta|}$$

$$E[\eta(t+\Delta)\eta(t)] = \sigma_\eta^2 \, e^{-k|\Delta|}$$

$$E[\xi(t)\eta(s)] = 0$$

which has some justification on physical grounds (see Section 4). (Higher order models have been considered -- see references cited.) A block-diagram of this engineering solution is shown in Figure 2.

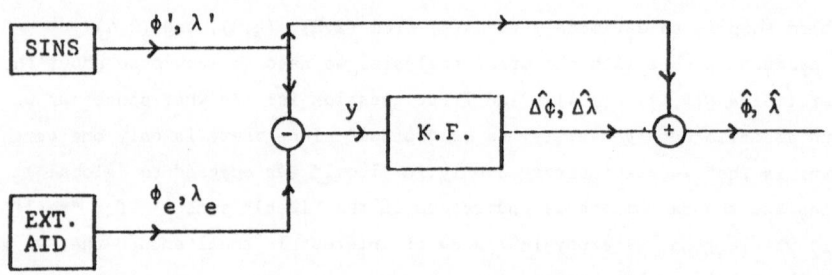

KALMAN FILTER MECHANIZATION

Figure 2

4. THE GENERAL PROBLEM

The engineering solution of the previous section was based on a grossly simplified model of the (time-domain) behavior of the V.D. field along the ships track. However, the general problem tends to be quite complex. First of all we need to have an independent measurement of the V.D. field $\xi(\phi,\lambda)$, $\eta(\phi,\lambda)$ along the ship's track but not along the true track since that is unknown. Thus let

$$\left.\begin{array}{l} \xi_o(t) = \xi(\phi'(t); \, \lambda'(t)) + N_5(t) \\ \eta_o(t) = \eta(\phi'(t); \, \lambda'(t)) + N_6(t) \end{array}\right\} \tag{4.1}$$

where $N_5(t)$, $N_6(t)$ represent the errors as a white noise or other process.

To make use of (4.1) we need to know more about the V.D. field. Here one may use the flat-earth version of the Vening-Meinesz formulas [5]. Thus it is known that the V.D. field can be expressed in terms of the gravity anomaly Δg. The gravity anomaly is defined as

$$\Delta g = g - \gamma$$

where γ is the "normal gravity field (magnitude) assuming an ellipsoidal body with the same mass as the earth, same center of mass and rotating with the same angular velocity and axis as the earth. Independent measurements of Δg (undulation of the geoid) are available, the most recent being by artificial satellites: satellite altimetry [9]. The Vening-Meinesz formulas are:

$$\xi(x,y) = \frac{1}{2\pi\gamma} \int_{-\infty}^{\infty} \int_{-\infty}^{\infty} \frac{\Delta g(u,v)(y-v)}{\left[(x-u)^2 + (y-v)^2\right]^{3/2}} \, dx \, dy \qquad (4.2)$$

$$\eta(x,y) = \frac{1}{2\pi\gamma} \int_{-\infty}^{\infty} \int_{-\infty}^{\infty} \frac{\Delta g(u,v)(x-u)}{\left[(x-u)^2 + (y-v)^2\right]^{3/2}} \, dx \, dy \qquad (4.3)$$

The properties of the V.D. field thus may be deduced from those of Δg. For example, if Δg is assumed to be a homogeneous isotropic Gaussian field, then

$$\sigma_\xi^2 = \sigma_\eta^2 = \frac{1}{2} \left. \sigma_{\Delta g}^2 \right| \gamma^2$$

and

$$E[\xi(x+\Delta x, \, y+\Delta y) \, \xi(x,y)] = \frac{1}{2} \frac{R_{\Delta g}(\Delta r)}{\gamma^2}$$
$$= E[\eta(x+\Delta x, \, y+\Delta y) \, \xi(x,y)]$$

where

$$r = \sqrt{\Delta x^2 + \Delta y^2} \ .$$

Many possible covariances for Δg have been suggested [1-4]. One of them is:

$$R_{\Delta g}(r) = \frac{C_0}{\left(1 + \dfrac{r^2}{D^2}\right)^\alpha} \qquad (4.4)$$

where $\alpha = 3/2$ or 2. One curious fact which is noteworthy is that the corresponding covariance for $\xi(vt, vt)$ is

$$R_\xi(t) = \frac{1}{2g^2} \frac{C_0}{\left(1 + \dfrac{t^2 v^2}{D^2}\right)^\alpha} \qquad (4.5)$$

which is not "realizable" in the sense that the corresponding spectral density does not satisfy the Wiener-Paley condition (is not "regular").

Getting back to our problem, if we use (4.4), (4.3), and (4.2), we do have a complete statement but the solution: finding the best mean-square estimate for $\Delta\phi$, $\Delta\lambda$ from (2.2), (2.3), and (4.1) still appears to be a formidable one. One possible simplification is to approximate $\xi(\phi,\lambda)$ by

$$\xi(\phi,\lambda) = \xi(\phi',\lambda') + \frac{\partial\xi}{\partial\phi} (\phi - \phi') + \frac{\partial\xi}{\partial\phi} (\lambda - \lambda')$$

and similarly for $\eta(\phi,\lambda)$. Here $\partial\xi/\partial\phi$, $\partial\xi/\partial\lambda$, $\partial\eta/\partial\phi$, and $\partial\eta/\partial\lambda$ are geodetic quantities and maps of these may be obtainable. However, substitution into (2.2) still does not reduce the problem to a standard one, so that the nonlinear filtering problem in the general case remains largely an open one.

5. ACKNOWLEDGEMENT

The author is indebted to Prof. W. Kaula for many helpful discussions on the geodesical aspects of the problem.

6. REFERENCES

1. W.M. Kaula, "Determination of the Earth's Gravitational Field," *Reviews of Geophysics*, Vol. 1, No. 4, 1963.

2. R.A. Nash and S.K. Jordan, "Statistical Geodesy -- An Engineering Perspective," *Proceedings of the IEEE*, Vol. 66, No. 5, May 1978.

3. W.M. Kaula, "Geodynamic Problems," *Proceedings of the 9th GEOP Conference, An International Symposium on the Applications of Geodesy to Geodynamics*, Ohio State University, Report No. 280, Columbus, OH, 1978.

4. S.K. Jordan, "Effects of Geodetic Uncertainties on a Damped Inertial System," *IEEE Transactions on Aerospace and Electronic Systems*, Vol. 9, No. 5, September 1973.

5. L. Shaw, I. Paul, and P. Henrikson, "Statistical Models for the Vertical Deflection from Gravity-Anomaly Models," *Journal of Geophysical Research*, Vol. 82, No. B7, 1978.

7. R.A. Nash, "Effect of Vertical Deflections and Ocean Currents on A Maneuvering Ship," *IEEE Transactions on Aerospace and Electronic Systems*, Vol. 4, No. 5, September 1968.

8. R.R. Hildebrant, K.R. Britting, and S.J. Madden, "The Effects of Gravitational Uncertainties on the Errors of Inertial Navigation Systems," *Journal of the Institute of Navigation*, Vol. 21, No. 4, Winter 1974-75.

9. W.D. Kahn, J.W. Siry, R.D. Brown, and W.T. Wells, "Ocean Gravity and Geoid Determination," *Journal of Geophysical Research*, Vol. 84, No. B8, 30 July 1979.

10. E.H. Metzger and A. Jircitano, "Inertial Navigation Performance Improvement Using Gravity Gradient Matching Techniques," *Journal of Spacecraft*, Vol. 13, No. 6, June 1976.

11. W.G. Heller and S.K. Jordan, "Error Analysis of Two New Gradiometer-Aided Inertial Navigation Systems," *Journal of Spacecraft*, Vol. 13, No. 6, June 1976.

12. P.S. Maybeck, *Stochastic Models, Estimation and Control*, Academic Press, 1979.

13. C. Broxmeyer, *Inertial Navigation Systems*, McGraw-Hill, 1964.

14. W.A. Heiskanen and H. Moritz, *Physical Geodesy*, W.H. Freeman & Co., 1967.

Research supported in part under Grant No. 83-0318, Applied Mathematics Division, AFOSR, United States Air Force.

FILTERING AND DETECTION PROBLEM FOR NONLINEAR TIME SERIES

Wojciech J. Kolodziej and Andrzej Pacut*
Department of Electrical & Computer Engineering
Oregon State University
Corvallis, OR 97331 USA

1. Introduction

Problem of filtering and detecting changes in dynamical properties of signals arises in various areas of control and signal processing. In particular, a number of methods is available to solve change detection problems [1], [2], [3]. An approach presented in this paper is to view a detection problem as a part of modeling discrete dynamical systems. Here we combine a state estimation algorithm for a system with non-Gaussian initial conditions with a change detection method. The abrupt system changes are modelled by the parameter changes of a dynamic model which is linear in unobservable variables but nonlinear in observable variables. The recursive finite dimensional filter presented here is derived for arbitrary distribution of the initial conditions and allows for a proper filter reinitialization after the parameter change. The filter output consists of the conditional probability density which is used by the detection algorithm. The detection method is based on the asymptotic local approach [4].

The methodology presented here is fairly general and bridges the model based signal processing with the statistical change detection and model validation techniques.

2. Model

Let us consider a nonlinear time series $\{x_t, \ t = 0,1,\ldots\}$ generated by the equations

$$x_{t+1} = f + Fx_t + Gw_t + Qv_t \ ,$$

$$y_{t+1} = h + Hx_t + v_t \ . \tag{1}$$

where $x \epsilon R^n$, $y \epsilon R^m$, $t \geqslant 0$; f, F, G, Q, h, H are known functions of an argument $(y_t, t; \theta)$, $Y_t = \{y_0, \ \ldots, \ y_t\}$, $\theta \epsilon R^p$. The initial condition x_0 is a random variable

---

* On leave from the Institute of Automatic Control, Technical University
  of Warsaw, Poland.

whose distribution function depends on an unknown parameter λ

$$P(a;\lambda) \overset{\Delta}{=} P(x_o \leqslant a | y_o;\lambda) , \qquad \lambda \epsilon R^q, \ a \epsilon R^n .$$

($x_o \leqslant a$ denotes n inequalities with respect to corresponding components of x_o and a) and y_o is a random variable. Sequences $\{w_t\}$ and $\{v_t\}$ are white, Gaussian, mutually independent and independent of x_o. Note that the above model includes a class of non-stationary stochastic models which are nonlinear in the measured data and accept arbitrary distribution of the initial condition.

Equations (1) are sufficiently general to serve as prototype models of various engineering problems. In particular, if (1) describes a system with unknown initial conditions $\lambda \epsilon R^q$, one may be interested in identification of the initial conditions distribution from the observations Y_t. Another example is a detection and identification problem. One can detect and identify the change in parameters $\theta \epsilon R^P$, i.e. to detect a structural change in the process statistical model.

A specific application of the latter is for nuclear power generation modeling. It is known that the neutron kinetics may be described by [5]:

$$d\eta = \left[\frac{1}{\tau} \left(\theta T_u - \beta \right)\eta + \alpha c + \frac{1}{2} \left(\frac{1-\beta}{\tau} \right)^2 \right]dt + \frac{1-\beta}{\tau} \eta dw_1 + sdw_2$$

$$dc = \left(\frac{\beta}{\tau} \eta - \alpha c \right)dt + dv_1 \qquad\qquad (2)$$

$$dT_u = \left(h\eta - uT_u \right)dt + dv_2$$

where

η – total number of neutrons

τ – average prompt neutron generation time

θ – temperature coefficient

α – average decay

c – average precursor population

β – average neutron generation rate by the precursors

s – external neutron source

T_u – core temperature

u – coolant mass flow rate

There are several sources of "randomness" in the above model: inaccurate measurements of the core temperature T_u, external neutron sources modeled by the additive noise dw_1, dw_2, dv_1, dv_2 and the average precursor population c, modeled usually as

$$c = \sum \gamma_i c_i$$

where c_i is a population of different precursors. An important detection problem for the above model is to detect the moment of a change in θ given measurement η (t_i), $i = 1,\ldots$. A discretized version of (2) leads to the model of the form (1).

3. Filtering

We will consider the filtering problem first. It may be shown [6] that a finite dimensional filter exists for the above model. Let $p(x_t|Y_t)$, $p(x_{t+1}|Y_{t+1})$, $p(x_{t+1}, y_{t+1}|Y_t)$ denote the corresponding conditional densities for the assumed values of λ and θ. Then the discrete version of the filter has a form

$$p(x_t|Y_t) = \frac{\int \phi(x_t; \bar{x}_t(a), \bar{P}_t)\psi_t(a)dP(a;\lambda)}{\int \psi_t(a)dP(a;\lambda)} \tag{3}$$

where

$\phi(\cdot\ ;\mu,\Sigma)$ denotes the Gaussian density with mean μ and variance Σ

$$\psi_t(a) = \exp\left(-\frac{1}{2}a^*S_t\,a + a^*R_t\right)\,,$$

$\bar{x}_t(a)$ and $\bar{P}_t$ are functions of observations Y_t to be specified later, and the superscript $*$ denotes transposition. The form (3) may be easily substantiated. Note that the conditional distribution $p(x_{t+1}, y_{t+1}|Y_t, x_t)$ of x_{t+1}, y_{t+1} given Y_t, x_t is Gaussian with expectation

$$m_t = \begin{bmatrix} f + F\,x_t \\ h + H\,x_t \end{bmatrix},$$

and covariance

$$\Sigma_t = \begin{bmatrix} QQ^* + GG^* & Q \\ Q^* & I \end{bmatrix}.$$

Consequently, if $p(x_t|Y_t)$ is assumed to have the form (3) then the density $p(x_{t+1}|Y_{t+1})$ is given by

$$p(x_{t+1}|Y_{t+1}) = \frac{\int p(x_{t+1}, y_{t+1}|Y_t, x_t)p(x_t|Y_t)dx_t}{\int p(y_{t+1}|Y_t, x_t)p(x_t|Y_t)dx_t} =$$

$$= \frac{\int\int [\phi(x_{t+1}, y_{t+1}; m_t, \Sigma_t)p(x_t; \bar{x}_t(a),\ \bar{P}_t)dx_t]\psi_t(a)dP(a;\lambda)}{\int\int [\phi(y_{t+1}; h+Hx_t, I)\phi\ (x_t; \bar{x}_t(a),\bar{P}_t)dx_t]\psi_t(a)dP(a;\lambda)}\ . \tag{4}$$

The above shows that $p(x_{t+1}|y_{t+1})$ has the form of equation (3). Certainly, (3) holds also for $t = 0$ and, consequently, it is true for all $t > 0$. The remaining functions $\bar{x}_t(a)$ and $\bar{P}_t$ can be calculated by the comparison of (3) and (4). After some algebra we can obtain the set of recurrent equations:

$$\bar{P}_{t+1} = (F\bar{P}_t F^* + GG^* + QQ^*) - (Q + F\bar{P}_t H^*)(I + H\bar{P}_t H^*)^{-1}(Q + F\bar{P}_t H^*)^* ,$$

$$\bar{P}_o = 0 ,$$

$$S_{t+1} = S_t + (H\Phi_t)^*(H\Phi_t) , \qquad\qquad S_o = 0 ,$$

$$R_{t+1} = R_t + \Phi^* H^*(y_{t+1} - h - H\bar{x}_t(0)) , \qquad\qquad R_o = 0 ,$$

$$\Phi_{t+1} = (F - (Q + F\bar{P}_t H^*)(I + H\bar{P}_t H^*)^{-1}H)\Phi_t , \qquad\qquad \Phi_o = I ,$$

$$\bar{x}_t(a) = \Phi_t a + \tilde{x}_t ,$$

$$\tilde{x}_{t+1} = f + F\tilde{x}_t + (Q + F\bar{P}_t H^*)(I + H\bar{P}_t H^*)^{-1}(y_{t+1} - h - H\tilde{x}_t)$$

$$\tilde{x}_o = 0$$

Note that the above equations have solutions dependent on Y_t (contrary to classical results in the linear-Gaussian case). The explicit solution of these equations is difficult to obtain except in some special cases like for the Gaussian mixture case, i.e. for $dP(a;\lambda) = \sum_i \lambda_i \phi(a;\mu_i,\Sigma_i)dx$.

4. Detection Problem

Now we consider a detection problem for (1). A practically useful detection algorithm should require little or no knowledge of the change time distribution. Sequential procedures are more appropriate. Let us assume the following model of the change of θ from θ_o to θ_1: until time t_{o-1} (inclusive)

$$p(y_{t+1}|Y_1^t) = p(y_{t+1}|Y_1^t,\theta_o) , \qquad\qquad t < t_o - 1 ,$$

and starting from $t > t_o$

$$p(y_{t+1}|Y_{t_o}^t) = p(y_{t+1}|Y_{t_o}^t,\theta_1) , \qquad\qquad t > t_o ,$$

where

$$Y_{t_o}^t = \{y_{to}, y_{to+1}, \ldots, y_t\} \, .$$

The value of θ prior to change at $t = t_o$ is known precisely (equal to θ_o) and it is completely unknown (equal to θ_1) after t_o.

It will be shown that with such a model the on-line detection problem is asymptotically equivalent to the sequential test of the following alternative hypotheses:

$$H_o : \quad \theta = \theta_o$$

$$\tag{5}$$

$$H_1 : \quad (\theta - \theta_o)^* F_t(\theta_o)(\theta - \theta_o) = \lambda_1^2 > 0 \, ,$$

where $F_t(\theta_o)$ is the Fisher information matrix and λ_1 is a parameter. To this end, we use a result of asymptotic expansion theory, which is based on the concepts of contiguity and weak convergence of a sequence of likelihood ratios. Let L_1^t denote a log-likelihood ratio

$$L_1^t(\theta_o, \theta) = \ln \frac{p(y_t | Y_1^{t-1}; \theta_o)}{p(y_t | Y_1^{t-1}; \theta)} \, , \qquad\qquad t > 1 \, .$$

If

$$\sqrt{t} \, (\theta - \theta_o) \xrightarrow[t \to \infty]{} \delta\theta$$

then [7] asymptotically, for $t \to \infty$

$$L_1^t(\theta_o, \theta) = \Delta_1^t(\theta_o) (\theta - \theta_o) - \frac{t}{2} (\theta - \theta_o)^* F_t(\theta_o) (\theta - \theta_o) + o(t) \, ,$$

where

$$\Delta^1(\theta_o) = \frac{\partial \ln p(y_t | Y_1^{t-1}; \theta_o)}{\partial \theta_o} \, ,$$

$$F_t(\theta_o) = E(\Delta_1^t(\theta_o)(\Delta_1^t(\theta_o))^*)$$

and $\dfrac{\Delta_1^t(\theta_o)}{\sqrt{t}}$ has asymptotically the normal distribution with the covariance matrix $F_t(\theta_o)$ and the expected value which depends on θ and is equal to

$$\begin{cases} 0 & \text{if } \theta = \theta_o \\[12pt] F_t(\theta_o)\delta\theta & \text{if } \theta = \theta_1 \end{cases}$$

From the above theorem it follows that the detection problem is asymptotically equivalent to a problem of testing the change-in-mean of the Gaussian sequence $\{\Delta_1^t\}$. The asymptotically sufficient statistic for the latter

$$\chi_t^2 = (\Delta_1^t(\theta_o))^* \, F_t^{-1}(\theta_o) \, \Delta_1^t(\theta_o) \, ,$$

has, for $\theta = \theta_o$, the non-central χ^2 distribution with p degrees of freedom and non-central parameter λ_1.

It is generally difficult to obtain $F_t(\theta_o)$. However, an appropriate sample covariance matrix $\hat{cov}$ may be used instead of the Fisher information matrix [8]. Consequently, a sequential detection algorithm may be based on a statistic:

$$L_{t-n_{t+1}}^t (\theta_o) \overset{\Delta}{=} \exp \left(-\frac{n_t \lambda_1^2}{2}\right) \, CHF \left(\frac{n_t}{\nu}, \frac{p}{2}, \frac{n_t \lambda_1^2 T_{n_t}^2 (\theta_o)}{2(n_{t-1}+T_{n_t}^2 (\theta_o))}\right)$$

where

$$CHF(a,b,\zeta) = 1 + \frac{a\xi}{b} + \frac{a(a+1)\zeta^2}{b(b+1)2!} + \dots \text{ is the confluent hypergeometric}$$

function,

$$T_{n_t}^2 (\theta_o) = \Delta_{t-n_{t+1}}^t (\theta_o)^* \, \hat{cov} \, (\Delta_{t-n_{t+1}}^t (\theta_o)) \, \Delta_{t-n_{t+1}}^t (\theta_o)$$

$$\Delta_{t-n_{t+1}}^t (\theta_o) = \sum_{k=t-n_{t+1}}^t \frac{\partial}{\partial \theta_o} \ln p(y_k|Y_{k-1};\theta_o) \, .$$

Let us define a decision function g_t as

$$g_t = \left(L_{t-n_{t+1}}^t (\theta_o) - \frac{\beta}{1-\alpha}\right)^+ , \tag{6}$$

where 1 is the unit-step function, the superscript + denotes a positive part, i.e. $x^+ = x1(x)$ and n_t is a counter since the last zero crossing of the decision function g_t, i.e.

$$n_t = n_{t-1} \, 1(g_{t-1}) + 1 \, . \tag{7}$$

The alarm time is now given as

$$t_a = \inf \{t: \ g_t > \frac{1-\beta}{\alpha} \} \, , \tag{8}$$

where α is the probability of false alarm, and β is the probability of delay. Since $L_{t-n_{t+1}}^{t}(\theta_o)$ is monotonic in $T_{n_t}^2(\theta_o)$, we can replace (6)-(8) with

$$g_t = (T_{n_t}^2(\theta_o) - T_L(n_t))^+ + T_L(n_t) ,$$

$$n_t = n_{t-1} \, 1(g_{t-1} - T_L(n_{t-1})) + 1 ,$$

$$t_a = \inf \{t: \; g_t > T_U(n_t)\}.$$

The upper $T_U(k)$ and the lower $T_L(k)$ thresholds can be pre-calculated from the relation

$$CHF \left(\frac{k}{2}, \frac{p}{2}, \frac{k\lambda_1^2 + T(k)}{2(k - 1 + T(k))}\right) = \begin{cases} \frac{\beta}{1-\alpha} , & T = T_L \\ \\ \frac{1-\beta}{\alpha} , & T = T_U . \end{cases}$$

Parameter λ_1 plays the role of a "dead" zone around the parameter θ_o and can be selected through a tuning procedure. This may be particularly useful if θ_o is unknown and is being recursively estimated from Y_t, with the corresponding estimation error $|\hat{\theta}_o - \theta_o|^2 \approx \lambda_1^2$.

The change time estimator $\hat{t}_o$ for the discussed algorithm is given by

$$\hat{t}_o = t_a - n_{t_a} + 1$$

We can define a time-window with the center at $\hat{t}_o$ and use the data within such a window to refine the above estimate (e.g., using the maximum likelihood method) and then apply an identification algorithm to estimate a new parameter value. The problem (5) can be generalized to yield:

$$H_o: \; (\theta-\theta_o)^* F_t(\theta_o) (\theta-\theta_o) = \lambda_o^2 ,$$

$$H_1: \; (\theta-\theta_o)^* F_t(\theta_o) (\theta-\theta_o) = \lambda_1^2 ,$$

$$\lambda_1 > \lambda_o .$$

The use of the "uncertainty" parameter λ_o enables the algorithm to compensate for the uncertainty of identified parameter θ_o.

The above procedure can be imbedded into a multi-level distributed detection structure. The first, lower level detects small changes in model parameters and adjusts a model as presented above. All information about these changes is sent to the upper level which may receive such signals from many lower-level sources. The upper level decides about major structural changes in the model. The information obtained by the upper level from the lower level observers may be fragmentary to decrease a required transmission channel capacity and increase the speed of data transmission. In particular, the information may just contain the moments of parameter adjustment. Because decisions about adjustments are based on likelihood ratios, these moments are actually the first passage process of the stochastic functions, through the levels equal to time-varying thresholds. This process again may be converted into a "slower" nonlinear ARMA process driven by a white noise.

5. References

[1]   M. Basseville and A. Benveniste, Eds., "Detection of Abrupt Changes in Signals and Dynamical Systems," Lec. Notes in Cont. and Inf. Sc., Vol. 77, New York, Springer-Verlag 1986.

[2]   A.S. Willsky, "A Survey of Design Methods for Failure Detection in Dynamics Systems," Automatica, Vol. 12, pp. 601-611, 1976.

[3]   A.V. Balakrishnan, "Minimal Time Detection of Parameter Change in a Counting Process," Lec. Notes in Cont. and Inf. Sc., ed. Archetti, Springer-Verlag, 1984.

[4]   A. Benveniste, M. Basseville, G.V. Moustakides, "The Asymplotic Local Approach to Change Detection and Model Validation," IEEE Trans. Aut. Cont., Vol. AC-32, No. 7, July 87.

[5]   R.R. Mohler, W.J. Kolodziej, "Stochastic Bilinear Models and Estimators with Nonlinear Observation Feedback," in U.B. Desai (Ed.) "Modelling and Application of Stochastic Processes," Kluver Acad. Publ., Boston 1986.

[6]   W.J. Kolodziej and R.R. Mohler, "State Estimation of Conditionally Linear Systems," SIAM J. Cont. Optimiz. 24, 497-508, 1986.

[7]   J. Deshayes and D. Picard, "Off-Line Statistical Analysis of Change-Point Models Using Non-Parametric and Likelihood Methods," in [1], 1986.

[8]   J.E. Jackson and R. Bradley, "Sequential χ^2- and T^2- Tests and their Application to an Acceptance Sampling Problem," Technometrics 3, N. 4, 1961.

SPECTRAL AND BISPECTRAL METHODS FOR THE ANALYSIS OF NONLINEAR
(NON GAUSSIAN) TIME SERIES SIGNALS

T. Subba Rao
Department of Mathematics
University of Manchester Institute of Science and Technology
PO Box 88
Manchester M60 1QD
U.K.

Abstract

In this paper we consider the estimation of spectrum and bispectrum of a
stationary time series. The usefulness of bispectrum to detect the periodicities of
signals when the signals are corrupted by noise is pointed out. We show that spectral
density function calculated from a fitted bilinear time series model can be a useful
alternative when the time series is nonlinear. We illustrate the estimation methods with
many simulated and one real data set. The real data is concerned with the magnetic
reversals.

1. Introduction

Spectral density function is the Fourier transform of autocovariance function of a
discrete parameter time series. The estimation of this function is an important part of
time series analysis. The need for power spectrum estimation arises in a variety of
contexts, such as measurement of noise spectra for the design of optimal linear filters,
the detection of narrow band signals, and the estimation of finite parameter linear
models. In recent years several methods of estimation of spectrum have been proposed,
and all these methods can be classified into one of the two categories, namely,
parametric or non parametric. The window estimation, which involves smoothing the
periodogram ordinates, is a non parametric method. The maximum entropy estimation (or
AR estimation) is a parametric method.

It is implicitly assumed in performing this analysis that the time series is linear
and perhaps Gaussian, and this assumption sometimes may be unrealistic.

The object of our paper is to propose methods of estimation of spectrum and
bispectrum of nonlinear time series. Since bilinear time series models (see Subba Rao,
1981) have second order properties similar to linear time series models, the spectral
density function can also be estimated from bilinear models. We propose a new method
of estimation of the bispectrum, and point out that it is always useful to estimate the
bispectrum also in addition to spectrum; especially when the observations are
corrupted by a Gaussian (coloured and white) noise. We illustrate these methods with
examples, and show how we can detect the periodicities in the signal.

We consider also the time series data concerned with the magnetic reversals, and
try to find the periodicities using the method described in the paper.

2. Spectral density function

Let $\{X(t)\}$ be a discrete parameter, real valued time series. We say that the time series $\{X(t)\}$ is second order stationary if

(i) $E(X(t)) = \mu$, independent of t

(ii) $\text{var}(X(t)) = \sigma^2_X$, independent of t   (2.1)

$\qquad \text{cov}(X(t)\ X(t+s)) = \gamma(s) = $ a function of the lag s only $(s = 0, \pm 1, \pm 2, \ldots)$.

It is well known that there exists a function $F(\omega)$, which is bounded and non decreasing such that

$$\gamma(s) = \int_{-\pi}^{\pi} e^{is\omega}\ dH(\omega) \tag{2.2}$$

The function $H(\omega)$ is known as the integrated (non-normalised) spectrum of the time series $\{X(t)\}$. If $H(\omega)$ is differentiable, $dH(\omega) = h(\omega)d\omega$, then (2.2) can be written as

$$\gamma(s) = \int_{-\pi}^{\pi} e^{is\omega}\ h(\omega)d\omega \tag{2.3}$$

The function $h(\omega)$ is known as the (non-normalised) spectral density function of the stationary time series $\{X(t)\}$.

From (2.3), we have

$$h(\omega) = \frac{1}{2\pi} \sum_{-\infty}^{\infty} \gamma(s)e^{-is\omega}. \quad |\omega| \leqslant \pi \tag{2.4}$$

If $\{X(t)\}$ is Gaussian, it is well known that all the information in the process $X(t)$ is contained in the mean and covariances $\{\gamma(s)\}$ and as such second order spectral analysis on the process $\{X(t)\}$ is sufficient to draw all the useful information about the process. If the process is not Gaussian, one may have to perform higher order spectral analysis, and in this paper we concentrate on bispectral density function and its use in the analysis of non Gaussian signals.

One of the objects in time series analysis is the estimation of $h(\omega)$, given a sample $(X(1), X(2),\ldots,X(n))$. There are several methods of estimation of $h(\omega)$. These methods can be divided into two categories, and they are (1) window estimates, which are non parametric (ii) parametric estimates. The window estimation consists of smoothing the periodogram ordinates with some standard windows, such as, Bartlett's, Tukey's, Parzen's window.

Briefly the window estimation can be described as follows.

Let $(X(1), X(2),\ldots,X(n))$ be a sample from $X(t)$. Obtain

$$\bar{X} = \frac{1}{n} \sum_{t=1}^{n} X(t), \quad \hat{\gamma}(s) = \frac{1}{n} \sum_{t=1}^{n-s} (X(t)-\bar{X})(X(t+|s|)-\bar{X}) \quad (s = 0,1,\ldots,(n-1)$$

$$\hat{h}(\omega) = \frac{1}{2\pi} \sum_{s=-(n-1)}^{n-1} \lambda\left(\frac{s}{M}\right) \hat{\gamma}(s) \cos \omega s \qquad (2.5)$$

where $M = M(n)$ and $\lambda(\cdot)$ is a lag window generator. If $\lambda(s) = 0$ for $|s| > 1$, M corresponds to the truncation point. We assume that the function $\lambda(s)$ is a bounded, even and square integrable such that $\lambda(0) = 1$. the integer M is chosen such that as $M \to \infty$, as $N \to \infty$ but $\frac{M}{N} \to 0$. It is well known that (see, e.g. Priestley (1981)).

$$E(\hat{h}(\omega)) \sim h(\omega) ,$$

$$\text{var}(\hat{h}(\omega) \sim \frac{M}{N} h^2(\omega) \int_{-\infty}^{\infty} \lambda^2(s) ds$$

which shows that $\hat{h}(\omega)$ is a consistent estimate of $h(\omega)$. The basic problem is to find the truncation point M and a suitable lag window. Several windows have been suggested in recent years, and the properties of these windows have been reviewed by Harris (1978) in the context of detection of harmonic signals in the presence of noise.

We now briefly consider AR spectral estimation. Suppose the time series $\{X(t)\}$ can be represented by the autoregressive model of order p, of the form

$$X(t) + a_1 X(t-1) + a_2 X(t-2) +\ldots+ a_p X(t-p) = e(t) \qquad (2.6)$$

where $\{e(t)\}$ is a sequence of uncorrelated random variables with mean zero and variance σ_e^2. Then we can show that $h(\omega)$ for the series $X(t)$ is given by

$$h(\omega) = \frac{\sigma_e^2}{2\pi|1+a_1 e^{-i\omega} + a_2 e^{-2i\omega+}\ldots+ a_p e^{-ip\omega}|^2} \qquad (2.7)$$

We note that when $X(t)$ satisfies (2.6), the covariance $\gamma(s)$ satisfies the Yule-Walker equations

$$\gamma(s) + a_1 \gamma(s-1)+\ldots+ a_p \gamma(s-p) = 0, \quad \text{if } s > 0 \qquad (2.8)$$

and therefore, one can obtain consistent estimates $(\hat{a}_1, \hat{a}_2,\ldots,\hat{a}_p)$ of $(a_1, a_2,\ldots,a_p)$ by solving the sample Yule Walker equations

$$\hat{\gamma}(s) + \hat{a}_1 \hat{\gamma}(s-1)+\ldots+ \hat{a}_p \hat{\gamma}(s-p) = 0, \quad (s = 1,2,\ldots,p) \qquad (2.9)$$

Then the AR spectral estimate is given by

$$\hat{h}(\omega) = \frac{\hat{\sigma}_e^2}{2\pi|1+\hat{a}_1 e^{-i\omega} + \hat{a}_2 e^{-2i\omega} +\ldots+ \hat{a}_p e^{-ip\omega}|^2} \qquad (2.10)$$

where $\hat{\sigma}_e^2$ is an unbiassed estimate of σ_e^2. The order (p) can be obtained by using Akaike's Information Criterion (AIC).

If the time series $\{X(t)\}$ is Gaussian, then the entropy is proportional to

$\int_{-\pi}^{\pi}$ log h(ω) dω. Maximising the entropy subject to some constraints leads to a spectral form similar to (2.10), and this equivalence has been established by Van den Bos (1971). The algorithm for developing maximum entropy spectral estimate was first given by Burg (1967). It must be noted that if the process {X(t)} is not Gaussian, no longer $\int_{-\pi}^{\pi}$ log h(ω)dω can be considered as an entropy, and therefore maximising this function is arbitrary, unless, one restricts to linear least squares predictors (and, therefore, to linear models) in which case the logarithm of innovation variance is proportional to $\int_{-\pi}^{\pi}$ log h(ω)dω.

This suggests that if the process is non-Gaussian, and one is interested to estimate the spectrum via a parametric model, it may be possible to estimate h(ω) through a nonlinear model which has second order properties similar to a linear model. One such nonlinear model is bilinear time series model (Granger and Andersen (1978), Subba Rao (1977, 1981), Subba Rao and Gabr (1984)). In this paper we will show that the estimation of h(ω) through a bilinear model is an alternative.

Another method which has received considerable attention in signal processing context was due to Pisarenko (1972, 1973). We will briefly discuss his approach for estimating h(ω), and later show how it can be extended to the estimation of bispectrum (for details see Subba Rao and Gabr (1986a, 1986b).

3. Pisarenko's method of estimation of h(ω)

Let {X(1), X(2),...,X(n)} be a sample of size n from a time series {X(t)} with E(X(t)) = 0. Let n = Mk, where M and k are integers. Divide the data into M groups, where each group consists of k observations, and let the observations in the ℓth group (ℓ = 1,2,...,M) be denoted by the vector $\underline{X}_\ell$, where

$$\underline{X}_\ell = (X((\ell-1)k+1), X((\ell-1)k+2),...,X(\ell k))$$
$$(\ell = 1,2,...,M)$$

For convenience, we shall label these observations as $X_\ell(1), X_\ell(2),...,X_\ell(k)$, where $X_\ell(m)$ = X((ℓ-1)k+m), (ℓ = 1,2,...,M; m = 1,2,...,k).

As an estimate of γ(t-s), consider

$$\hat{R}_k(t,s) = \frac{1}{M} \sum_{j=1}^{M} X_j(t) X_j(s) \qquad (t,s = 1,2,...,k)$$

and the kxk covariance matrix $\hat{\underline{R}}_k$, where

$$\hat{\underline{R}}_k = \frac{1}{M} \sum_{j=1}^{M} \underline{X}_j \underline{X}_j' \qquad (3.2)$$

Since $E(\hat{R}_k(t,s)) \simeq R_k(t,s)$, $E(\hat{\underline{R}}_k) = \underline{R}_k$. Let $\hat{\lambda}_{k,j}$ $(j = 0,1,2,...,n-1)$ be the eigen values of $\hat{\underline{R}}_k$, let the corresponding normalised eigen vectors be $\hat{\underline{a}}_{k,j}$ $(j = 0,1,2,...,n-1)$, where

$$\hat{\underline{a}}'_{k,j} = (\hat{\underline{a}}_{k,j}(0), \hat{\underline{a}}_{k,j}(1),...,\hat{a}_{k,j} \ (k-1))$$

Consider a spectral estimate of $h_k(\omega)$ of the form

$$\hat{h}_k(\omega_\varrho) = \frac{1}{2\pi k} \sum_{j=0}^{k-1} \hat{\lambda}_{k,j} \ |\hat{B}_{k,j}(\omega_\varrho)|^2 \tag{3.3}$$

where

$$\hat{B}_{k,j}(\omega_\varrho) = \sum_{t=0}^{k-1} \hat{a}_{k,j}(t) e^{it\omega_\varrho} \tag{3.4}$$

In fact, $\hat{h}_k(\omega_\varrho)$ can also be written as

$$\hat{h}_k(\omega_\varrho) = \frac{1}{2\pi k} \sum_{t=0}^{k-1} \sum_{s=0}^{k-1} \hat{R}_k(t,s) \ e^{-i(t-s)\omega_\varrho} \tag{3.5}$$

which is Bartlett's   estimate. The advantage of writing $\hat{h}_k(\omega_\varrho)$ in terms of the eigen values $\{\hat{\lambda}_{k,j}\}$ is that one can replace $\hat{\lambda}_{k,j}$ by any nonlinear (continuous) function, say $G(\hat{\lambda}_{k,j})$ and obtain the Pisarenko's(1972) estimate

$$\hat{h}_k(\omega_\varrho) = g\left[\sum_{j=0}^{k-1} G\left[\hat{\lambda}_{k,j}\right] |\hat{B}_{kj}(\omega_\varrho)|^2 \right] \tag{3.6}$$

where g is an inverse function of $G(\cdot)$ such that $g[G(x)] = x$. The high resolution estimator of Capon (1969) is obtained by setting $G(x) = x^{-1}$. Using the properties of circulant symmetric matrices, in recent papers Subba Rao and Gabr (1986a,b) obtained the asymptotic properties of the above estimates. Briefly, they are as follows. If $\hat{h}_k(\omega_\varrho)$ is given by (3.3), then

$$E(\hat{h}_k(\omega_\varrho)) \simeq h_k(\omega_\varrho) + 0(\frac{1}{M}) \ ,$$

$$\text{var}(\hat{h}_k(\omega_\varrho)) \simeq \begin{cases} \dfrac{h^2_k(\omega_\varrho)}{M} & \text{if} \quad \omega_\varrho \neq 0, \pi \\[3mm] \dfrac{2h^2_k(\omega_\varrho)}{M} & \text{if} \quad \omega_\varrho = 0, \pi \ , \end{cases} \tag{3.7}$$

and the estimates $\hat{h}_k(\omega_\varrho)$ and $\hat{h}_k(\omega_m)$, $(\omega_\varrho \neq \omega_m)$ are approximately uncorrelated.

4. Bilinear models

Bilinear models have been introduced by control theorists (see Brockett, (1976), Mohler (1973), Ruberti et al.(1972)), and these models have been found to be very useful in describing many nonlinear phenomena. These models are "nearly" linear, and their structural properties are very similar to the linear models. In the context of time series, these models have been introduced by Granger and Andersen (1978a, 1978b), Subba Rao (1977, 1981a, 1981b) and the properties of these models and fitting of these models to real and simulated data have been discussed in detail by Subba Rao and Gabr (1984). Briefly, we summarise some results relevant to the present paper.

Let {X(t)} be a discrete parameter time series, satisfying the difference equation

$$X(t) + \sum_{j=1}^{p} a_j X(t-j) = \sum_{j=0}^{r} c_j e(t-j) + \sum_{\ell=1}^{m} \sum_{\ell'=1}^{k} b_{\ell\ell'} X(t-\ell) e(t-\ell') \qquad (4.1)$$

where {e(t)} is a Gaussian white noise process. The above model is represented as BL(p,r,m,k). Consider the bilinear model BL(p,0,p,1); and this can be written in matrix notation as follows;

$$\underline{x}(t) = \underline{A}\ \underline{x}(t-1) + \underline{B}\ \underline{x}(t-1)\ e(t-1) + \underline{C}\ e(t)\ ,$$
$$X(t) = \underline{H}\ \underline{x}(t) \qquad (4.2)$$

where

$$\underline{A} = \begin{bmatrix} -a_1 & -a_2 & \cdots & -a_p \\ 1 & 0 & \cdots & 0 \\ & & \vdots & \\ 0 & 0 & \cdots & 0 \end{bmatrix} \quad \underline{B} = \begin{bmatrix} b_{11} & b_{21} & b_{31} & \cdots & b_{p1} \\ 0 & 0 & 0 & \cdots & 0 \\ & & \vdots & & \\ 0 & 0 & 0 & \cdots & 0 \end{bmatrix} \qquad (4.3)$$

$\underline{C}' = (1,0,\ldots 0)$, $\underline{H}' = (1,0,\ldots 0)$ and $\underline{x}'(t) = (X(t), X(t-1),\ldots,X(t-p+1)$

To study some features of bilinear time series we have generated time series {X(t)} from the model

$$X(t) = 0.4\ X(t-1) + 0.8\ X(t-1)\ e(t-1) + e(t) \qquad (t = 1,2,\ldots,1000)$$

and the plot is given in Fig.1. An examination of the plot shows that at certain time points, there are high amplitude oscillations; and these could be interpreted as "outliers" because such high amplitude oscillations usually cannot occur with stationary linear models. Robust spectral estimation techniques recently developed by Kleiner, Martin and Thompson (1979) (see also the book edited by Franke, Härdle and Martin (1984)) are aimed at estimating the spectral density function of time series with such features. As an alternative to the robust estimation, we can fit a bilinear model to such data (because nonlinear models, in particular, bilinear models can produce such features) and then calculate the spectrum from these fitted models, and this is possible as we can obtain the theoretical expressions for tthe covariances for bilinear models. Here we derive these results for the model BL(p,0,p,1).

Let us assume that {e(t)} are independent, and each is distributed normally with mean zero and variance unity.

Let $\underline{C}(s) = E(\underline{x}(t+s)-\underline{\mu})(\underline{x}(t)-\underline{\mu})'$, then we can show that (see Subba Rao, 1981)

$$\underline{C}(0) = \underline{A}\,\underline{C}(0)\,\underline{A}' + \underline{B}\,\underline{C}(0)\,\underline{B}' + \Delta_2$$

$$\underline{C}(1) = \underline{A}\,\underline{C}(0) + \Delta_3, \qquad\qquad\qquad (4.4)$$

$$\underline{C}(s) = \underline{A}\,C(s-1) = \underline{A}^{s-1}C(1), \quad (s = 2,3,\ldots)$$

where

$$\underline{\mu} = E(\underline{X}(t)),$$

$$\Delta_2 = \underline{B}\,\underline{\mu}\,\underline{\mu}'\underline{B}' + \underline{A}\,\underline{\mu}\,\underline{\mu}'\underline{A}' + \underline{A}\,\underline{S}\,\underline{B}' + \underline{B}\,\underline{S}\,\underline{A}'$$
$$\quad + 2\underline{B}\,\underline{C}\,\underline{C}'\,\underline{A}' + \underline{C}\,\underline{C}' - \underline{\mu}\,\underline{\mu}',$$

$$\Delta_3 = \underline{A}\,\underline{\mu}\,\underline{\mu}' + \underline{B}\,\underline{S} - \underline{\mu}\,\underline{\mu}',$$

$$\underline{S} = \underline{A}\,\underline{\mu}\,\underline{C}' + \underline{B}\,\underline{C}\,\underline{C}' + \underline{C}\,\underline{\mu}'\underline{A}' + \underline{C}\,\underline{C}\,\underline{B}',$$

when $\underline{A}$ and $\underline{B}$ are of the form (4.3), from (4.4) we have the difference equation

$$\gamma(s) + a_1\,\gamma(s-1) + a_2\,\gamma(s-2) + \ldots + a_p\,\gamma(s-p) = 0, \quad s > 1$$

The above equations are the same as the Yule-Walker equations for ARMA(p,1) and thus show that the bilinear model BL(p,0,p,1) has the same covariance structure as an ARMA(p,1). Therefore, one has to compute higher order moments in order to distinguish linear time series models and bilinear time series models. Recently, Sesay and Subba Rao (1986) have shown that bilinear models of the above kind satify difference equations in third order moments and fourth order moments (cumulents).

We can now give an expression for the parametric spectral density function of the model BL(p,0,p,1). First, we calculate the spectral density matrix of the vector process $\underline{x}(t)$, which we shall denote by $\underline{h}_x(\omega)$. From the definition of the spectral matrix, we have

$$\underline{h}_x(\omega) = \frac{1}{2\pi} \sum_{s=-\infty}^{\infty} \underline{C}(s)\,e^{-is\omega} = \frac{1}{2\pi}\left[\underline{C}(0) + \underline{M}\,\underline{C}(1) + \underline{C}(1)\,\underline{M}^*\right] \qquad (4.5)$$

where $\underline{M} = e^{i\omega}\,(I - \underline{A}\,e^{-i\omega})^{-1}$, $\underline{M}^*$ is the complex conjugate of $\underline{M}$. We use 'vec' notation, where vec($\underline{D}$) is

$$\text{vec}(\underline{D}) = \begin{bmatrix} \underline{D}_{\cdot 1} \\ \underline{D}_{\cdot 2} \\ \underline{D}_{\cdot p} \end{bmatrix},$$

$\underline{D}_{\cdot 1}, \underline{D}_{\cdot 2}, \ldots, \underline{D}_{\cdot p}$ are the column vectors of the pxp matrix D.

With this notation, we have (see Sesay, 1985)

$$\text{vec}(\underline{h}_x(\omega)) = \frac{1}{2\pi}\left\{ [\underline{I} + \underline{I}\otimes(\underline{M}\underline{A}) + (\underline{M}^*\,\underline{A})\otimes I]\,[I - \underline{A}\otimes\underline{A} - \underline{B}\otimes\underline{B}]^{-1}\,\text{vec}(\Delta_2)\right.$$

$$\left. + [(I\otimes M) + (\underline{M}^*\otimes I)]\,\text{vec}(\Delta_3) \right\},$$

$\underline{A} \circledast \underline{A}$ are Kronecker products.

The first element of $vec(\underline{h}_x(\omega))$ is the expression for the spectral density function of the bilinear model.

5. Bispectral density function

Suppose we have a stationary (up to third order) time series having the one sided moving average representation

$$X(t) = \sum_{u=0}^{\infty} g(u) \, e(t-u), \quad g(0) = 1, \tag{5.1}$$

where $\{e(t)\}$ are independent, with $E(e(t)) = 0$, $E(e^2(t)) = \sigma^2$, $E(e^3(t)) = \mu_3$. Then we can show that the spectral density function $h(\omega)$ of $X(t)$, given by (5.1), is

$$h(\omega) = \frac{\sigma_e^2}{2\pi} \, |G(e^{-i\omega})|^2 \tag{5.2}$$

where

$$G(e^{-i\omega}) = \sum_{u=0}^{\infty} g(u) \, e^{-iu\omega}$$

The bispectral density function (which may be complex valued), $h(\omega_1, \omega_2)$, is defined as Fourier transform of the third order covariances given by

$$h(\omega_1, \omega_2) = \frac{1}{(2\pi)^2} \sum_{\tau_1=-\infty}^{\infty} \sum_{\tau_2=-\infty}^{\infty} C(\tau_1, \tau_2) \, e^{-i\tau_1\omega_1 - i\tau_2\omega_2} \tag{5.3}$$

$$-\pi \leqslant \omega_1, \, \omega_2 \leqslant \pi$$

where $C(\tau_1, \tau_2) = E(X(t)-\mu)(X(t+\tau_1)-\mu)(X(t+\tau_2)-\mu)$

$$= C(\tau_2, \tau_1) = C(-\tau_1, \tau_2-\tau_1) = C(\tau_1-\tau_2, -\tau_2).$$

In view of the above symmetry relations for $C(\tau_1, \tau_2)$, we have

$$h(\omega_1, \omega_2) = h(\omega_2, \omega_1) = h(-\omega_1, -\omega_1-\omega_2) = h(-\omega_1-\omega_2, \omega_2) = h(-\omega_1, -\omega_2) \tag{5.4}$$

Because of symmetry, it is sufficient to calculate $h(\omega_1, \omega_2)$ for (ω_1, ω_2) defined in section (2) of Fig.2.

If $X(t)$ satisfies (5.1), then we can show that

$$C(\tau_1, \tau_2) = \mu_3 \sum_{r} g(r) \, g(r+\tau_1) \, g(r+\tau_2) \, ,$$

$$\tag{5.5}$$

$$h(\omega_1, \omega_2) = \frac{\mu_3}{(2\pi)^2} \, G(e^{-i\omega_1}) \, G(e^{-i\omega_2}) \, G(e^{i\omega_1+i\omega_2})$$

If $\mu_3 = 0$, $h(\omega_1, \omega_2) = 0$ for all ω_1 and ω_2. For Gaussian processes $\mu_3 = 0$; and therefore, the bispectrum is zero. However, there may exist a process X(t) which is non Gaussian, but admitting a linear representation of the form (5.1), then from the relation (5.5), we obtain the ratio

$$\frac{|h(\omega_1,\omega_2)|^2}{h(\omega_1)\ h(\omega_2)\ h(\omega_1+\omega_2)} = \text{constant} \tag{5.6}$$

In other words, bispectral density function can be used to detect any departure from linearity (and Gaussianity) of the signal process; and this is the basis of Subba Rao and Gabr's test for linearity of time series (Subba Rao and Gabr, 1980).

In later sections, we will illustrate with examples the usefulness of bispectrum for detecting the periodicities of the signal in the presence of noise. We now consider a realistic situation where we do not observe the signal X(t), but observe a contaminated signal Z(t), where for each time t, we have

$$Z(t) = X(t) + Y(t) \tag{5.7}$$

where the noise Y(t) is a zero mean, third order stationary process independent of X(t). We can easily show that

$$E(Z(t)) = 0,$$

$$R_Z(s) = E(Z(t)Z(t+s)) = R_X(s) + R_Y(s),$$

$$h_Z(\omega) = h_X(\omega) + h_y(\omega), \tag{5.8}$$

$$C_Z(\tau_1,\tau_2) = E(Z(t)Z(t+\tau_1)Z(t+\tau_2))$$

$$= C_X(\tau_1,\tau_2) + C_Y(\tau_1,\tau_2),$$

$$h_Z(\omega_1,\omega_2) = h_X(\omega_1,\omega_2) + h_Y(\omega_1,\omega_2) \tag{5.9}$$

Now suppose we wish to estimate the parameters of the signal {X(t)} (for example, they may correspond to the frequencies of the signal) given the covariances {$\hat{R}_Z(s)$} (or equivalently $\hat{h}_Z(\omega)$; the efficiency of these estimates do depend on the knowledge of the signal to noise ratio, and the bandwidths of the noise spectrum $h_y(\omega)$ and the signal spectrum $h_X(\omega)$. If it happens that the stationary noise process {Y(t)} has pseudo periods, and if they lie within the range of the frequencies of the signal {X(t)}, it becomes almost impossible to distinguish the genuine peaks of the signals from the pseudo peaks of the signal. Therefore the usual assumption that is often made is that the noise {Y(t)} is a white noise with spectrum $h_Y(\omega) = \frac{\sigma_y^2}{2\pi}$, and the signal X(t) is a is a finite linear model, say an autoregressive model of order p of the form (2.6). Then it is well known that the covariance structure of Z(t) is similar to that of an ARMA(p,p), with the AR coefficients the same as AR coefficients of (2.6). This similarity has been exploited by Pagano (1974) for estimating the AR coefficients and the variance of the noise; and Subba Rao (1976) has shown that the model can be written in the form of a linear state space form; and the dimension of the state is equal to the order of the model. Using the analogy with canonical factor analysis proposed by Rao

(1965, p.456), Subba Rao has shown that it is sufficient to perform a prinicipal component analysis for the prediction errors of the observed process {Z(t)}, to estimate not only the order of the model, but also the parameters of the AR model and the variance of the noise. As shown by Ulrych and Clayton (1976), the algorithm proposed by Pisarenko (1973) for estimating the harmonics of the signal does in fact depend on performing the principal component analysis on the observed vector {Z(t)}. We will discuss this further in later sections.

We observe from (5.8) that the problem of identifying the frequencies of the signal heavily depends on the noise structure, even if we assume that {Y(t)} is a Gaussian white noise. If the assumption of symmetry of the distribution can be made, then we see from that (5.9),

$$h_Z(\omega_1,\omega_2) = h_X(\omega_1,\omega_2).$$

Therefore, the third order spectrum (bispectrum) of the observed process is the same as the bispectrum of the signal, and hence any important frequencies of the signal can be estimated immediately from the bispectrum. These estimates are more robust against the presence of coloured noise, and these ideas will be illustrated in section 6.

6. Estimation of the Bispectrum

There are several methods of estimation of bispectrum, and two widely used methods are (i) using fast Fourier transforms and complex demodulation techniques (see Huber et al., 1971; Godfrey, 1963) (ii) window technique (see Subba Rao and Gabr (1984)). Here we briefly discuss an alternative method proposed by Subba Rao and Gabr (1986b) and based on an extension of Pisarenko's method.

Let (X(1), X(2),...,X(n)) be a sample from a zero mean third order stationary process {X(t)}. Let $J_X(\omega)$ be the finite Fourier transform

$$J_X(\omega) = \sum_{t=1}^{n} X(t)e^{it\omega} ,$$

and let $I_n(\omega_1,\omega_2,\omega_3)$ given by

$$I_n(\omega_1,\omega_2,\omega_3) = \frac{1}{n(2\pi)^2} J_X(\omega_1) J_X(\omega_2) J_X(\omega_3)$$

be the third order periodogram.

Then we can show that, if $\omega_1+\omega_2+\omega_3 = 0 \pmod{2\pi}$,

$$h_n(\omega_1,\omega_2) = E(I_n(\omega_1,\omega_2,\omega_3))$$

$$= \frac{1}{n(2\pi)^2} \sum_{t_1} \sum_{t_2} \sum_{t_3} C(t_1-t_3,t_2-t_3)e^{-i(t_1-t_3)\omega_1-i(t_2-t_3)\omega_2}$$

$$= \frac{1}{n(2\pi)^2} \sum_{s_1=1}^{2n-1} \sum_{s_2=1}^{2n-1} C^*(s_1-n, s_2-n) \, e^{-i(s_1-n)\omega_1 - i(s_2-n)\omega_2} \tag{6.1}$$

where

$$C^*(s_1,s_2) = \begin{cases} C^*(s_2,s_1) & \text{if } 0 \leqslant s_2 \leqslant s_1 \leqslant n-1 \\ & \text{(i.e.) if } (s_1,s_2) \text{ belongs to sector (2)} \\ (n-\max(s_1,s_2))C(s_1,s_2) & \text{if } 0 \leqslant s_1 \leqslant s_2 \leqslant s_1 \leqslant n-1 \\ & \text{(i.e.) if } (s_1,s_2) \text{ belongs to sector (1)} \\ C^*(s_2-s_1,-s_1) & \text{if } (s_1,s_2) \text{ belongs to sector (3)} \\ C^*(s_1-s_2,-s_2) & \text{if } (s_1,s_2) \text{ belongs to sector (4)} \\ C^*(-s_2,s_1-s_2) & \text{if } (s_1,s_2) \text{ belongs to sector (5)} \\ C^*(-s_1,s_2-s_1) & \text{if } (s_1,s_2) \text{ belongs to sector (6)} \\ 0 & \text{otherwise} \end{cases}$$

(The sectors defined above correspond to Fig. 1.2 of Subba Rao and Gabr (1984, p.13.)

Define the symmetric matrix $\underline{C}^*$ of order $(2n-1) \times (2n-1)$,

$$\begin{bmatrix} C^*(n-1,n-1) & C^*(n-1,n-2) & \cdots & C^*(n-1),0) \\ C^*(n-2,n-1) & C^*(n-2,n-2) & \cdots & C^*(n-2,0) & C^*(n-2,-1) & 0 & 0 & \cdots & 0 \\ C^*(0,n-1) & C^*(0,n-2) & \cdots & C^*(0,0) & C^*(0,-1) & \cdots & & & C^*(0,-n+1) \\ 0 & C^*(-1,n-2) & \cdots & C^*(-1,0) & C^*(-1,-1) & \cdots & & & C^*(-1,-n+1) \\ 0 & 0 & & C^*(-n+1,0) & C^*(-n+1,-1) & \cdots & & & C^*(-n+1,-n+1) \end{bmatrix}$$

Let $\{\mu_{n,j}, j = -(n-1),\ldots 0,1,\ldots(n-1)\}$ be the eigen values of $\underline{C}^*$ and let $\underline{C}_{n,-(n-1)}$, $\underline{C}_{n,-(n-2)}\cdots \underline{C}_{n,(n-1)}$ be the corresponding normalised eigen vectors.

Since $\underline{C}^*$ is symmetric, we have the representation

$$\underline{C}^* = \sum_{j=-(n-1)}^{n-1} \mu_{n,j} \, \underline{C}_{n,j} \, \underline{C}'_{n,j} \tag{6.2}$$

where

$$\underline{C}'_{n,j} = (C_j(-n+1) \, C_j(-n+2),\ldots,C_j(n-1)).$$

Substituting for $C^*(s_1-n,s_2-n)$ from (6.2) into (6.1), we get

$$h_n(\omega_1,\omega_2) = \frac{1}{(2\pi)^2 n} \sum_{j=-(n-1)}^{n-1} \mu_{n,j} \, \underline{C}_{n,j}^*(\omega_1) \, \underline{C}_{n,j}^*(\omega_2) \tag{6.3}$$

where

$$c_{n,j}^*(\omega) = \sum_{s=-(n-1)}^{n-1} c_j(s)e^{-is\omega}.$$

In the following, we estimate $h_n(\omega_1,\omega_2)$ given by (6.3).

As in section (3), we group the n observations $(X(1),X(2),...X(n))$ into M groups, each group having k observations. We estimate the third order moments by

$$\hat{C}_j(s_1,s_2) = \frac{1}{M} \sum_{\ell=1}^{M} (X_\ell(j)-\overline{X})(X_\ell(j+s_1)-\overline{X}_{j+s_1})(X_\ell(j+s_1)-\overline{X}_{j+s_2}),$$

and the smoothed estimate by

$$\hat{c}_j^*(s_1,s_2) = \frac{1}{k} \sum_{j=1}^{k-\tau} \hat{C}_j(s_1,s_2), \quad (s_1 = 0,\pm 1,\pm 2,...,\pm(k-1), \ s_2 = 0,\pm 1,...,\pm(k-1) \tag{6.4}$$

where $\tau = \max(s_1,s_2)$.

The third order covariances $\tilde{C}_j(s_1,s_2)$ are calculated over (s_1,s_2) defined in sector (2).

Let $\{\hat{\mu}_{k,j}, \ j = 0,\pm 1,...,\pm(k-1)\}$ be the eigen values and $\{\hat{\underline{C}}_{kj}\}$ be the corresponding eigen vectors of $\hat{\underline{C}}_k$. Then the bispectral estimate is given by

$$\hat{h}_k(\omega_1,\omega_2) = \frac{1}{k(2\pi)^2} \sum_{j=-k+1}^{k-1} \hat{\mu}_{k,j} \ c_{k,j}^*(\omega_1) \ c_{k,j}^* (\omega_2) \tag{6.5}$$

where

$$c_{k,j}^*(\omega) = \sum_{s=-(k-1)}^{k-1} c_{k,j}(s)e^{-is\omega}.$$

In the estimation of a second order spectral density $h(\omega)$, we have seen earlier that the eigen values can be replaced by any nonlinear function, and by suitably defining an inverse function one can recover the spectral estimate. This suggestion of Pisarenko (1973) can be justified (see Subba Rao and Gabr (1986a)) using the properties of the circulant symmetric matrices $\underline{R}_k$. Here in the case of bispectrum, we note that $\hat{\underline{C}}_k$ is not a circulant symmetric matrix, and therefore such an interpretation is not possible. Still, we can write the bispectral estimate in terms of the eigen values and the eigen vectors of $\hat{\underline{C}}_k$. The advantage of writing in terms of the eigen values is that a visual inspection of the absolute values give an indication of the number of eigen values to be included in the summation (6.5) and this is not so if we estimate the bispectrum using the third order covariances.

7. The estimation of the periodicities through spectrum and bispectrum

In order to understand why AR spectrum can be used to estimate the periodicities of the signal, let us assume that the signal X(t) satisfies the model

$$X(t) = \sum_{j=1}^{m} A_j \sin(\omega_j t + \phi_j) \qquad (7.1)$$

where $\{A_j\}$ are amplitudes and $\{\phi_j\}$ are phases. It can be shown (see Chan et al. (1981)) that there exists a unique real 2m vector $(a_1, a_2, ..., a_{2m})$ such that X(t) can be written in the form of a difference equation

$$X(t) = \sum_{j=1}^{2m} a_j X(t-j) \qquad (7.2)$$

where $\{a_j\}$ satisfy the equation

$$\sum_{j=1}^{2m} a_j \cos \omega_i j = 0$$

$$\sum_{j=1}^{2m} a_j \sin \omega_i j = 0 \qquad (7.3)$$

In other words, X(t) can be considered as an autoregressive model of order 2m with excitation force being zero. The zeros of the polynomial $Z^{2m}-a_1 Z^{2m-1}...-a_{2m}=0$ give the frequencies $(\omega_1, \omega_2, ..., \omega_m)$ (see Kay and Marple, 1981).

In real situations, we do not observe the signal X(t), but observe a contaiminated version, namely, Z(t) = X(t) + Y(t), where Y(t) is the noise process. In this case, Z(t) is not an AR process, but instead an ARMA(2m,2m), and as pointed out earlier the methodology must be somewhat different if one wants to use a parametric approach (see Clayton and Ulrych (1976), Subba Rao (1976)).

Here we give some examples, to illustrate the use of spectrum and bispectrum for estimating the periodicities (for more examples see Subba Rao and Gabr, 1986b).

Example 1

A time series $\{X(t)\}$ is generated from the model

$$Z(t) = 2 \sin \frac{\pi}{4} t + e(t) \qquad (7.4)$$

where $\{e(t)\}$ are independent, identically distributed random normal variables with mean zero and variance unity. The parameters chosen for the estimation are n=3000, k=30 and M=100. The spectrum and the bispectrum are estimated using the relations (3.3) and (6.5) respectively. In figures (3) and (4), we plot the spectrum and the modulas of the bispectrum. In the spectrum we see a clear peak at $\omega=0.25\pi$, and in the modulas of the bispectrum at $\omega_1 = \omega_2 = 0.25\pi$.

Example 2

A time series {X(t)} is generated from the model

$$Z(t) = 4 \sin (0.15\pi)t + 4 \sin (0.35\pi)t + e(t) \tag{7.5}$$

For estimating the spectrum the parameters chosen are k=50, n=100 and n=5000; and for the bispectrum the parameters chosen are n=3000, k=30 and M=100. The spectrum and the modulas of the bispectra are given in Figures (5) and (6). The peaks are clearly visible in both graphs at the frequencies 0.15π and 0.35π.

Example 3

This example shows the usefulness of the bispectrum.

The time series {Z(t)} is generated from the model

$$Z(t) = 4 \sin(0.15\pi)t + 4 \sin(0.55\pi)t + Y(t), \tag{7.6}$$

where Y(t) is a coloured noise satisfying the model

$$Y(t) - 0.4 \ Y(t-1) + 0.7 \ Y(t-2) = e(t) \tag{7.7}$$

where {e(t)} is defined as in Example 1. We note that the spectrum of $Y(t)$, $h_y(\omega)$, has a peak at 0.4π which is in between the frequencies 0.15π and 0.55π of the signal X(t). The spectral estimate $\hat{h}_k(\omega)$ is calculated using the parameters

(i) n = 5000, k = 50 and M = 100   (ii) n = 400, k = 40 and M = 100

The graphs of the spectra are given in Figures (7,8). In the case of k = 50 and M = 100, we observe clear peaks at the frequencies at ω = 0.15π and 0.55π corresponding to the signal, whereas in the case k = 40 and M = 100, there are no visible peaks at these frequencies. Instead we observe a peak at ω = 0.4π corresponding to the noise.

Since {Y(t)} is Gaussian, we have $h_Z(\omega_1,\omega_2)$ = $h_X(\omega_1,\omega_2)$, and therefore, we should have clear peaks at the frequencies of the signal in the modulas of the bispectrum. We calculated the bispectral estimate using k = 30 and M = 120. The modulas of the bispectra is plotted in Fig.(9). We see clear peaks of ω_1 = ω_2 = 0.15π, ω_1 = ω_2 = 0.55π and small peaks at ω_1 = 0.15π, ω_2 = 0 and ω_2 = 0.55π, ω_1 = 0. No significant peak is observed at ω = 0.4π corresponding to the noise {Y(t)}.

This example clearly indicates the usefulness of bispectrum in detecting the periodicities of the signal. The author believes that it is always useful to calculate both the spectrum and the bispectrum, and if peaks appear in both at the same frequencies, one can conclude that they are genuine. Besides, if the bispectrum is non zero, it also indicates that the time series is non Gaussian, and even nonlinear.

8. The spectral and bispectral analysis of the earth's magnetic reversals

It is well known that reversals of the earth's magnetic field may occur with a certain regularity. The theoretical results postulate long term periodicity in magnetic stratigraphy with reversal periods of 285, 114, 64, 47 and 34 million years. Recently several authors(Stothers (1986), Negi and Tiwari (1983), Raup (1985), Lutz (1985)), have analysed this data for detecting for the presence of regular cycles. Negi and Tiwari (1983) have based their analysis on Walsh spectrum and have come to the conclusion that spectral peaks around 285, 114, 64, 47 or 34 million years seem to be very significant. Since the data corresponds to reversals over intervals of time, Walsh analysis is no doubt more appropriate.

The data sets analysed by various authors seem to be different, and hence there could be some inconsistency in their conclusions. Recently, Stothers (1986) considered the 296 magnetic reversals over the past 165 million years, the dates (intervals) of these reversals are given by Harland et al. (1982). The data analysed by Stother (1986) corresponds to the number of reversals over 4 million year intervals.

In this paper, we considered the number of reversals during the first 128 million years as given by Harland (1982); the data corresponding to the number of reversals over 2 million year intervals. Thus we have 62 observations, the spacing between successive observations is 2 million years. The number of observations we are analysing is quite short to draw any firm conclusions; and we are presenting the analysis to emphasize the methodology described in this paper and other recent papers by the author. Still, we are encouraged to find that the conclusions we have drawn are quite close to the observations made by the geophysicists. We hope to analyse in detail, if and when more data sets are made available to use. The plot of the data is given in Fig.(10), and the spectrum and the modulas of the bispectrum with k=6, M=10 are given in Figures (11) and (12). We see that the data is highly skew, possibly has a negative exponential distribution. The bispectral estimate is quite large confirming that the series is non Gaussian.

Bilinear model of the form,

$$X(t) = 1.27 \ X(t-1) - 0.32 \ X(t-2) - 0.026 \ X(t-1) \ e(t-1) - 0.135 \ X(t-2) \ e(t-1) + e(t),$$

where {e(t)} are independent, with estimated variance 2.0937 of the residuals fitted better than a linear ARIMA model. The spectrum obtained from this model is very smooth, and no cler peaks are visible in the spectrum. Most of the power is in the low frequency band, which may be an indicator of the presence of "long periodicity" in the data.

In the computed periodogram, we observed a significant peak at the frequency corresponds to period 64, and smaller peaks at 8 and 5 units. In view of the spacing of 2 million years in the data, the above corresponds to periods 128, 16 and 10 million years. The period 128 is near to 120 which corresponds to the periodicity of the interaction of two arm spiral density wave with sun's galactic orbit (see Negi and Tewari, 1983). The periods 16 and 10 have been observed also by Stothers (1986).

Now let us consider the spectral estimate and the bispectral estimate obtained by choosing k=6 and M=10. We do not see any clear peak in the low frequency of the spectrum (see Fig. (11)) but observe two small peaks at frequency $\omega=0.45\pi$ (corresponding to approximately 9 m years) and another at $\omega=0.8\pi$ (approximately 5m yr).

It is instructive to see the values of the modulas of the bispectrum, and therefore these are given in Table (1). We see that the values are very large in the low frequency range bands, confirming that this might be due to long periodicity in the data. The value at $\omega_1 = \omega_2 = 0.05\pi$ is very significant. Though this does not correspond to a peak, we see that there is a sudden drop in the value at the next frequency calculated. Had we calculated the bispectrum at "finer" grid of frequencies using long data sets, we would have observed significant peaks in the neighbourhood of this frequency. Of course, this frequency corresponds to 80 million years, and as pointed out by Negi and Tiwari (1093) this may correspond to the variational period of sun perpendicular to the galactic plane which is 85m. years. There are other periods at $\omega_1=0$, $\omega_2=0.45\pi$ corresponding to, nearly, 9m. years and the other at $\omega_1=0$, $\omega_2=0.8\pi$ corresponding to 5m. years for periodicity. These peaks are observed by Stothers (1986) and others.

The above preliminary analysis shows that bispectrum can be very useful in detecting for the presence of periodicities in series which are non Gaussian and which are possibly corrupted by additive independent Gaussian noise.

Acknowledgements: The spectrum and bispectrum are calculated using the computer programs written by Dr. M.M. Gabr, and a listing of these can be obtained from the author on request. I am thankful to Mrs. B.K. Stensholt for performing the calculations reported in this paper.

References

Brockett, R.W. (1976) Volterra series and geometric control theory. Automatica, 12, 167-172.

Burg, J.P. (1967) Maximum entropy spectral analysis. Paper presented at the 37th annual international S.E.G. meeting, Oklahoma City, Oklahoma.

Capon, J. (1969) High resolution frequency - wave number spectrum analysis. Proc. I.E.E.E. 57, 1408-1418.

Chan, Y.T. and Lavoie, J.M.M. and Plant, J.B. (1981) A paramertic estimation approach to estimation of frequencies of sinusoids. I.E.E.E. Acoustics, speech and signal processing, ASSP-29, No.2, 214-219.

Franke, J., Härdle, W. and Martin, D. (1984) Robust and nonlinear time series. Lecture notes in Statistics, 26, Springer-Verlag, New York.

Godfrey, M.D. (1965) An exploratory study of the bispectrum of economic time series. J. Roy. Statis. Soc. C 14, 48-69.

Granger, C.W.J. and Andersen, A.P. (1978a) An introduction to bilinear time series models. Vandenhoeck and Ruprecht, Göttingen.

Granger, C.W.J. and Andersen, A.P. (1978b) Nonlinear time series modelling in applied time series analysis, ed. by D.F. Findley, Academic Press.

Harland, W.B. (1982) A geological time scale. Cambridge Univ. Press.

Harris, F.J. (1978) On the use of windows for harmonic analysis with discrete Fourier transform. Proc. I.E.E.E. Vol. 66, No.1, 51-83.

Huber, P. J., Kleiner, B. and Gasser, T. (1971) Statistical methods for investigating phase relations in stationary stochastic processes. I.E.E.E. Trans. Audio and Electroacoustics. Vol. A U-19-20.

Kay, S.M. and Marple, S.L. (1981) Spectrum analysis. A modern perspective. Proc. I.E.E.E. 69, No. 11, 1380-1419.

Kleiner, B., Martin, R.D., and Thomson, D.J. (1979) Robust estimation of power spectra, J. Roy. Statist. Soc., Series B, 41, 313-351.

Lutz, T.M. (1985) The magnetic reversal record is not periodic. Nature 317, 404-407.

Mohler, R.R. (1973) Bilinear control processes. Academic Press.

Negi, J.G. and Tiwari, R.K. (1983) Matching long term periodicities of geomagnetic reversals and galactic motions of the solar system. Geophysical Research letters, 10. No. 8, 713-716.

Pagano, M. (1974) Estimation of models of autoregressive signal plus white noise. Annals of Statistics, 2, 99-108.

Pisarenko, V.F. (1972) On the estimation of spectra by means of nonlinear functions of the covariance matrix. Geophysical J. Roy. Astronomical Soc. 28, 511-531.

Pisarenko, V.F. (1973) The retrieval of harmonics from a covariance function. Geophysical J. Roy. Astronomical Soc. 33, 347-366.

Priestley, M.B. (1981) Spectral Analysis and Time Series. Academic Press.

Raup, D.M. (1985) Rise and fall of periodicity. Nature 317, 384-385.

Rao, C.R. (1967) Linear statistical inference and its applications. John Wiley, New York.

Sesay, S.A.O. (1985) Frequency domain methods of estimation and higher order moment analysis for the bilinear model BL(p,o,p,1). Unpublished Ph.D. thesis submitted to the University of Manchester.

Sesay, S.A.O. and Subba Rao, T. (1986) Yule-Walker type difference equations for higher order moments (and cumulents) for bilinear time series models. (Submitted for publication).

Subba Rao,T. (1976) Canonical factor analysis and stationary time series models. SankhyaSer. B. 38, 256-271.

Subba Rao, T. (1977) On the estimtion of bilinear time series models. Bull. Internatinal. Statist. Inst. 41.

Subba Rao, T. and Gabr, M.M. (1980) A test for linearity of stationary time series. J. Time Ser. Anal. 1, 145-158.

Subba Rao, T. (1981) On the theory of bilinear time series models. J.R.
    Statist. Soc. B 43, 244–255.

Subba Rao, T. and Gabr, M.M. (1984) An introduction to bispectral
    analysis and bilinear time series models. Lecture notes in Statistics,
    No. 24, Springer Verlag.

Subba Rao, T. and Gabr, M.M. (1986a) Estimation of spectrum, inverse
    spectrum and bispectrum of a stationary time series – I. (Under
    preparation).

Subba Rao, T. and Gabr, M.M. (1986b) Estimation of spectrum, inverse spectrum
    and bispectrum of a stationary time series – II. (Under preparation).

Stothers, R.B. (1986) Periodicity ofthe earth's magnetic reversals. Nature
    444–446.

Ulrych, T.J. and Clayton, R.W. (1976) Time Series modelling and maximum entropy.
    Physics of the earth and planetary interiors 12, 188–200.

Van den Bos, A., (1971) Alternative interpretation of maximum entropy
    spectral analysis. I.E.E.E. Trans. Inf; Th. It-17,493-494.

Table 1 — Values of $|h(\omega_1,\omega_2)|$, with $j \uparrow\ (\omega_j = j\pi)$ (rows) and $\rightarrow j\ (\omega_j = j\pi)$ (columns).

j	.00	.05	.10	.15	.20	.25	.30	.35	.40	.45	.50	.55	.60	.65
1.00	.094													
.95	.323	.138												
.90	.861	.494	.212											
.85	1.364	1.025	.181	.270										
.80	1.524	1.362	.519	.165	.259									
.75	1.283	1.310	.907	.391	.192	.054								
.70	.875	.960	1.027	.578	.096	.032	.250							
.65	.676	.635	.830	.538	.185	.012	.208	.515						
.60	.937	.654	.520	.347	.220	.019	.111	.362	.685					
.55	1.583	1.108	.394	.252	.168	.055	.044	.157	.577	.611				
.50	2.234	1.775	.586	.284	.181	.101	.051	.051	.326	.410	.441			
.45	2.446	2.222	1.080	.483	.223	.124	.032	.020	.102	.151	.347	.174		
.40	2.049	2.127	1.585	.859	.225	.127	.126	.201	.019	.083	.143	.072	.033	
.35	1.376	1.576	1.746	1.122	.359	.165	.229	.412	.150	.032	.127	.120	.023	.090
.30	1.201	1.145	1.484	1.135	.560	.242	.267	.520	.425	.251	.157	.201	.069	
.25	2.382	1.504	1.200	1.147	.712	.374	.217	.456	.637	.522	.051			
.20	5.371	3.374	1.266	1.382	.941	.644	.169	.264	.649					
.15	9.856	6.983	1.851	1.516	1.313	1.012	.348							
.10	14.743	11.607	3.998	1.997	1.564									
.05	18.5391	15.928	7.596											
.00	19.970													

Table 1   Values of $|h(\omega_1,\omega_2)|$

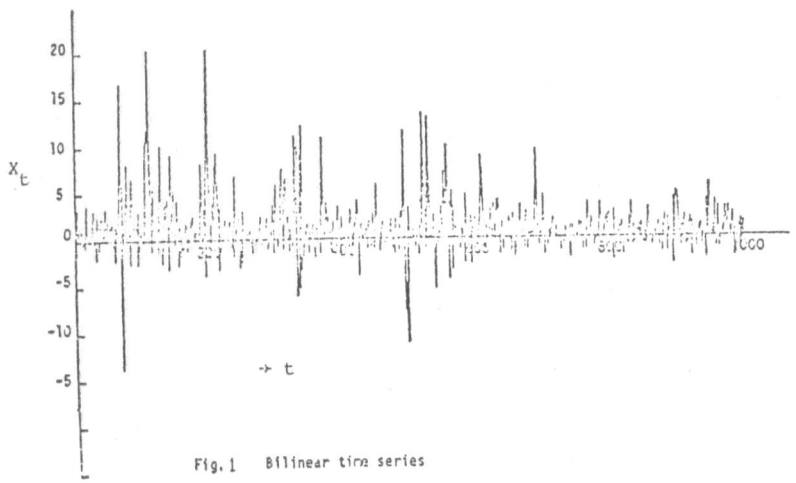

Fig. 1   Bilinear time series

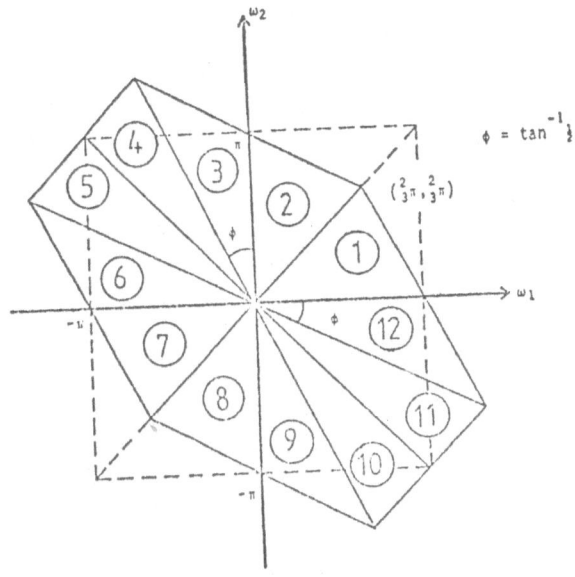

Fig. 2   Regions of $h(\omega_1, \omega_2)$

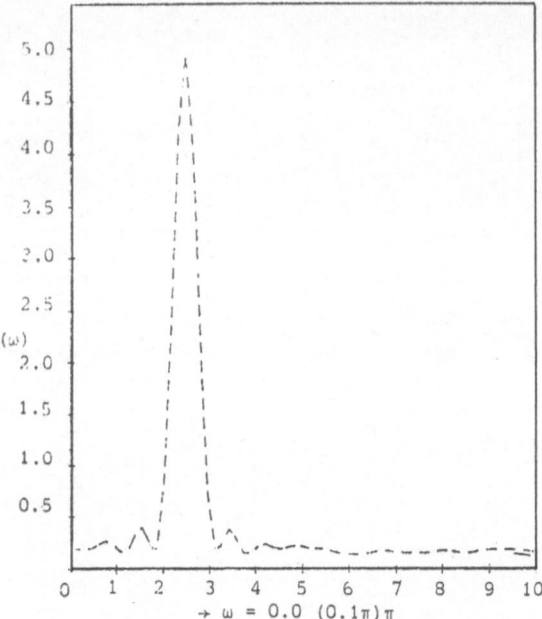

Fig. 3 Spectrum of the series (2.4)

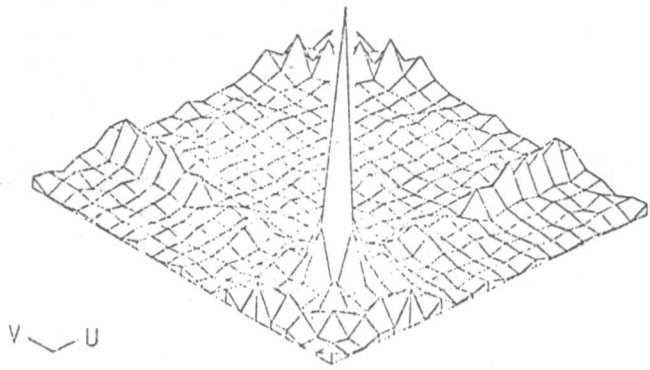

Fig. 4 Modulus of the bispectrum

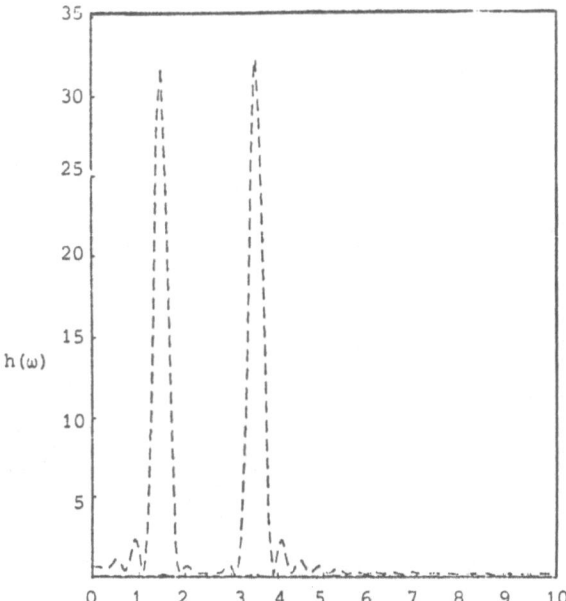

Fig. 5 Spectrum of the series (7.3)

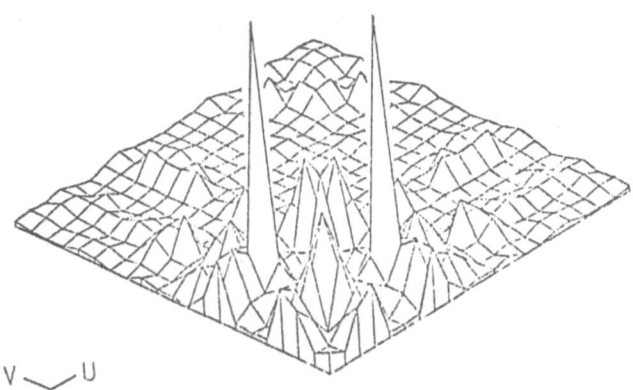

V ‿ U

Fig. 6 Modulus of the bispectrum

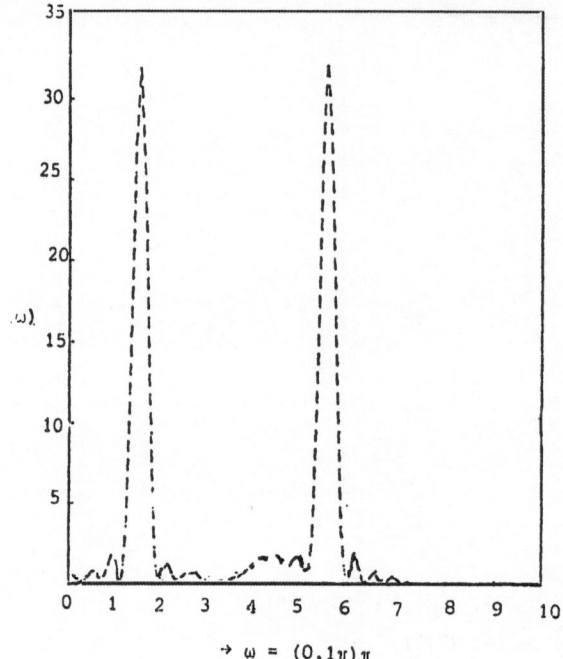

Fig. 7 Spectrum of the series (7.6)

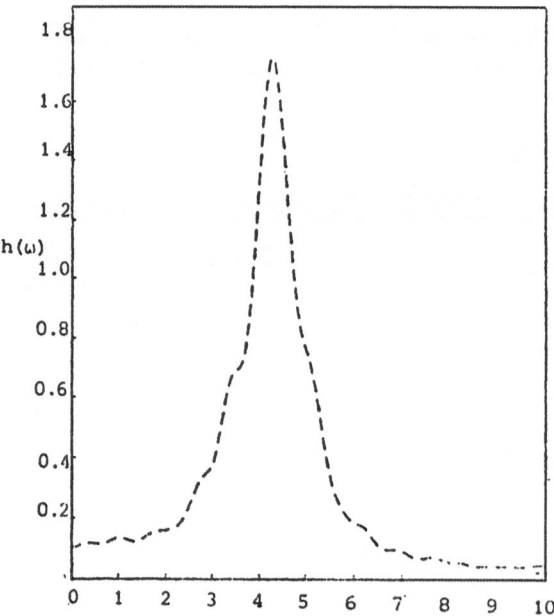

Fig. 8 Spectrum of the series (7.6)

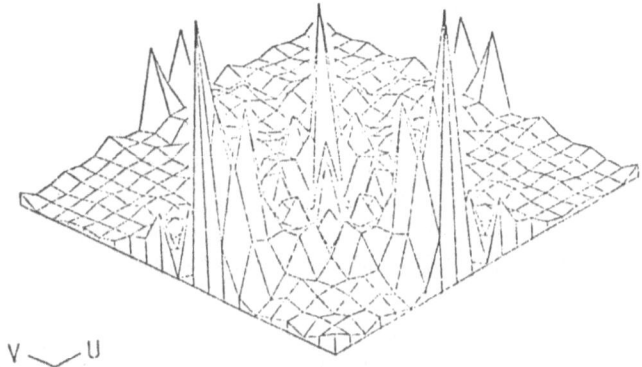

V ⌣ U

Fig. 9 Modulus of the bispectra of (7.6)

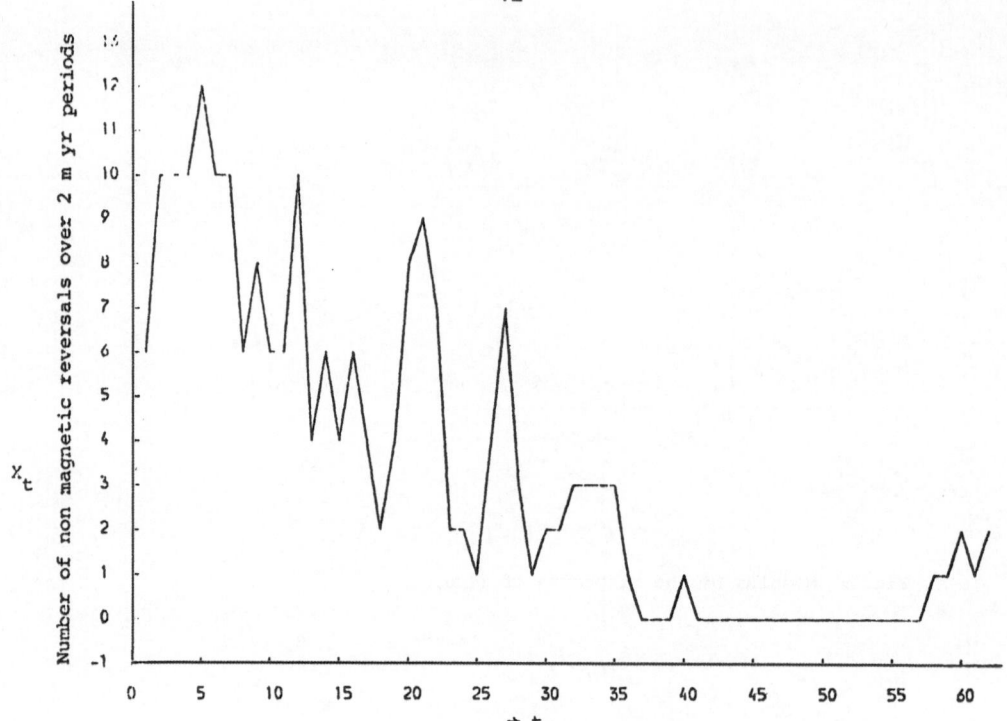

Fig. 10 Magnetic Reversals

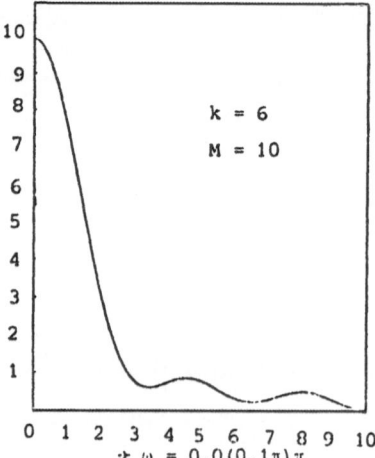

k = 6

M = 10

ω = 0.0(0.1π)π

Fig. 11 Spectrum of the Magnetic Reversals

BILINEAR TIME SERIES: THEORY AND APPLICATION

Z. Tang and R.R. Mohler
Department of Electrical & Computer Engineering
Oregon State University
Corvallis, OR 97331 USA

1.   INTRODUCTION

The purpose of this chapter is to: a) overview the analysis of the bilinear
time series, BTS; b) to show its application to real-world processes; and c) to
provide some results for the affine BTS, ABTS. Conditions for degrees of stationarity and ergodicity are studied along with the application to the parameter-
estimation of a scalar model. A vector ABTS is shown to include the realization of
non-anticipative affine, bilinear, autoregressive, moving-average models, AB(p,p) of
order p, as a special case.

Linear time-series models, including ARMA models, have been popular for the
statistical characterization of Gaussian random processes and dynamic systems for a
number of years. In many applications, however, the assumptions of linearity and/or
Gaussianity are not valid. For example, acoustical signal processing may involve
such phenomena as a consequence of random variations in the physical properties of
the conducting media. This may include shallow bay temperature variations from
seasonal currents. The overall effect of sound-speed inhomogeneities at a given
signal frequency is concisely given in terms of a medium "strength parameter" and
"diffraction parameter" in the analysis of wave propagation through turbulent media
[1]. Depending on the region, the media amplitude distortion may be approximately
lognormal, Raleigh, or Rice-Nakagami [1,2]. In particular, lognormal and Raleigh
distributions are easily characterized by the proposed methods.

Research on bilinear systems originated from control studies for nuclear
fission, heat transfer, socio-economics and biomedicine [3], and apparently BTS
analysis can be quite useful in these applications. A couple of BTS examples are
analyzed in Section 2 below. Bilinear models may evolve naturally or as a better
approximation to more highly nonlinear processes. In the late 1970's, research on
BTS evolved in Manchester under T.S. Rao and M.B. Priestley. The publications in
this and related areas (e.g., see [3-22]) form a good base for continued research in
this area. Hinich [23], Rao and Gabr [24] successfully tested data for Gaussianity
and linearity by estimating their third-order cumulants and their bispectrum. These
tests fit very nicely when it is necessary to adapt to changing data structures.
Various degrees of stationarity and ergodicity for BTS have been tested in the
literature. Obviously, the more rigid the requirement, the more limited the

admissible parameter values of the BTS. These aspects are studied in Sections 4 and 5 below.

Here the affine bilinear time series, ABTS, is defined by the following:

$$X_k = AX_{k-1} + BX_{k-1}u_{k-1} + C_0u_k + C_1u_{k-1} + D , \tag{1}$$

where k is the index of time; $X_k \in R^n$, u_k is a standard, white, Gaussian scalar sequence with zero mean and unity variance which is independent of X_0; and A, B, C_0, C_1, D are constant matrices of appropriate dimension. If D is zero, of course (1) is the traditional BTS.

2. APPLICATIONS

BTS and ABTS arise naturally in numerous cases and are valid approximations in certain others as noted above. The structure of part of the *human immune system* has been modeled as a connection of bilinear processes which evolve naturally [25,26]. Cell population dynamics in the immune system (as well as other population models) may sometimes be derived by

$$x_k^i = v_{k-1}^i - \frac{x_{k-1}^i}{\tau_i} + p_{k-1}^i x_{k-1}^i + \sum_{j \neq i} 2p_{k-1}^j p_{k-1}^{ji} x_{k-1}^j - \sum_{\ell \neq i} 2p_{k-1}^\ell p_{k-1}^{i\ell} x_\ell , \tag{2}$$

where x_k^i is the population or concentration of the ith class of cells; v_k^i is the source of cells (e.g., bone marrow via blood); τ_i is death time constant; p_k^i, p_k^{ji}, $p_k^{i\ell}$ are appropriate growth coefficients (random processes) which include "probabilities" of stimulation and differentiation from one class to another. In immunology, i may be indexed according to different classes of T cells, B cells, macrophages, natural killers, mast cells, and tumor cells. The theory of estimation of such models is studied in [26,27].

The following nonlinear model of a constant-speed, point object which moves in a plane with variable heading is a good example of how a BTS may arise in the discrete-time estimation of parameters for a continuous process. The continuous motion is described by

$$d(\sin\phi)/dt = (\cos\phi)(d\phi/dt) ,$$
$$d(\cos\phi)/dt = -(\sin\phi)(d\phi/dt) ,$$
$$dy/dt = v(\sin\phi) ,$$
$$dx/dt = v(\cos\phi) . \tag{3}$$

Here, ϕ is the heading angle; x,y are the rectangular coordinates, and v is velocity magnitude. Suppose the angular velocity is given by

$$d\phi/dt = \omega + e_t , \tag{4}$$

where ω is a constant, and e_t is a standard white Gaussian process with zero mean and unit intensity. Choosing the state vector

$$z_t^T = (\sin\phi, \cos\phi, y, x) , \tag{5}$$

$$dz_t/dt = vAz_t + Bz_t(d\phi/dt) , \tag{6}$$

where

$$A = \begin{pmatrix} 0 & 0 \\ \bar{A} & 0 \end{pmatrix} , \quad B = \begin{pmatrix} \bar{B} & 0 \\ 0 & 0 \end{pmatrix} , \quad \bar{A} = \begin{pmatrix} 1 & 0 \\ 0 & 1 \end{pmatrix} , \quad \bar{B} = \begin{pmatrix} 1 & 1 \\ -1 & 0 \end{pmatrix} .$$

Here z_t is a stochastic process, which satisfies the stochastic differential equation (Ito sense)

$$dz_t = \left(vA + \omega B + \tfrac{1}{2} \sigma^2 B^2 \right) z_t dt + \sigma B z_t dw_t , \tag{7}$$

where w_t is a standard Wiener process.
Then, it can be shown that

$$z_t = \begin{pmatrix} \mathrm{Exp}(B\omega t)\mathrm{Exp}(B\sigma w_t) & 0 \\ v \int_0^t \mathrm{Exp}(B\omega r)\mathrm{Exp}(B\sigma w_t)dr & I_2 \end{pmatrix} z_0 ; \tag{8}$$

where I_2 is the 2x2 identity matrix.

For estimating the mean of the angular velocity ω discrete data is used, and the Euler method is used for discrete approximation. If the sampling time δt is small enough the accuracy can be guaranteed for a fixed-time interval. The discrete approximation of (7) is

$$z_{k+1} - z_k = \left(vA + \omega B + \tfrac{1}{2} \sigma^2 B^2 \right) \delta t z_k + \sigma \sqrt{\delta t} B z_k u_k , \tag{9}$$

where u_k is an identically Gaussian-distributed independent random sequence with zero mean and unity variance; z_k is independent of $u_k, u_{k+1}, \ldots$. Usually only coordinates x,y are measurable, and it is convenient to let $z_k^T = (z_{1k}^T, z_{2k}^T)$. Then

$$z_{1,k+1} = \left(I + \omega\delta t B - \tfrac{1}{2}\sigma^2\delta t I_2 \right) z_{1,k} + \sigma\sqrt{\delta t}Bz_{1,k}u_k \; ,$$

$$z_{2,k+1} = z_{2,k} + v\delta t z_{1,k} \; . \tag{10}$$

Here $z_{1,k}$, $z_{2,k}$, $z_{2,k+1}$ are independent of u_k, $u_{k+1},\dots$. From (10),

$$z_{2,k+2} = z_{2,k+1} + v\delta t\left(\left(1 - \tfrac{1}{2}\sigma^2\delta t\right)I_2 + \omega\delta t B \right)z_{1,k} + v\sigma\delta t^{3/2}Bz_{1,k}u_k$$

$$= z_{2,k+1} + \left[\left(1 - \tfrac{1}{2}\sigma^2\delta t\right)I_2 + \omega\delta t B \right](z_{2,k+1} - z_{2,k})$$

$$+ \sigma\sqrt{\delta t}B(z_{2,k+1} - z_{2,k})u_k \; . \tag{11}$$

From given data $z_{2,k}$, for each k let

$$\epsilon_k = z_{2,k+2} - z_{2,k+1} - \left[\left(1 - \tfrac{1}{2}\sigma^2\delta t\right)I_2 + \omega\delta t B \right](z_{2,k+1} - z_{2,k})$$

$$+ \sigma\sqrt{\delta t}B(z_{2,k+1} - z_{2,k})u_k$$

$$= (z_{2,k+2} - 2z_{2,k+1} + z_{2,k}) + \tfrac{1}{2}\sigma^2\delta t(z_{2,k+1} - z_{2,k})$$

$$- \omega\delta t B(z_{2,k+1} - z_{2,k}) - \sigma\sqrt{\delta t}(z_{2,k+1} - z_{2,k})u_k \; .$$

ϵ_k is random due to the measurement error, and it is independent of u_k. The estimator of ω, $\hat{\omega}$, is selected so that $E\left[\displaystyle\sum_{k=-n}^{n} \epsilon_k^T\epsilon_k \right]$ is a minimum. Then, null

$$\partial E\left[\sum_{k=-n}^{n} \epsilon_k^T\epsilon_k \right]\Big/\partial\omega\Big|_{\omega=\hat{\omega}} = 2\hat{\omega}\delta t \sum_{k=-n}^{n} (z_{2,k+1} - z_{2,k})^T B^T B(z_{2,k+1} - z_{2,k})$$

$$- \tfrac{1}{2}\sigma^2\delta t^2\left\{ \sum_{k=-n}^{n} (z_{2,k+1} - z_{2,k})^T B(z_{2,k+1} - z_{2,k}) + \sum_{k=-n}^{n} (z_{2,k+1} - z_{2,k})^T B^T(z_{2,k+1} - z_{2,k}) \right\}$$

$$-\delta t\left\{ \sum_{k=-n}^{n} (z_{2,k+2} - z_{2,k+1} + z_{2,k})^T B(z_{2,k+1} - z_{2,k}) + \sum_{k=-n}^{n} (z_{2,k+1} - z_{2,k})^T B^T(z_{2,k+2} - z_{2,k+1} + z_{2,k}) \right\}$$

Since $B^T = -B$,

$$0 = \partial E\left[\sum_{k=-n}^{n} \epsilon_k^T\epsilon_k \right]\Big/\partial\omega\Big|_{\omega=\hat{\omega}}$$

$$= 2\hat{\omega}\delta t \sum_{k=-n}^{n} (z_{2,k+1} - z_{2,k})^T B^T B(z_{2,k+1} - z_{2,k})$$

$$- 2\delta t\left\{ \sum_{k=-n}^{n} (z_{2,k+2} - z_{2,k+1})^T B(z_{2,k+1} - z_{2,k}) \right\} \; .$$

47

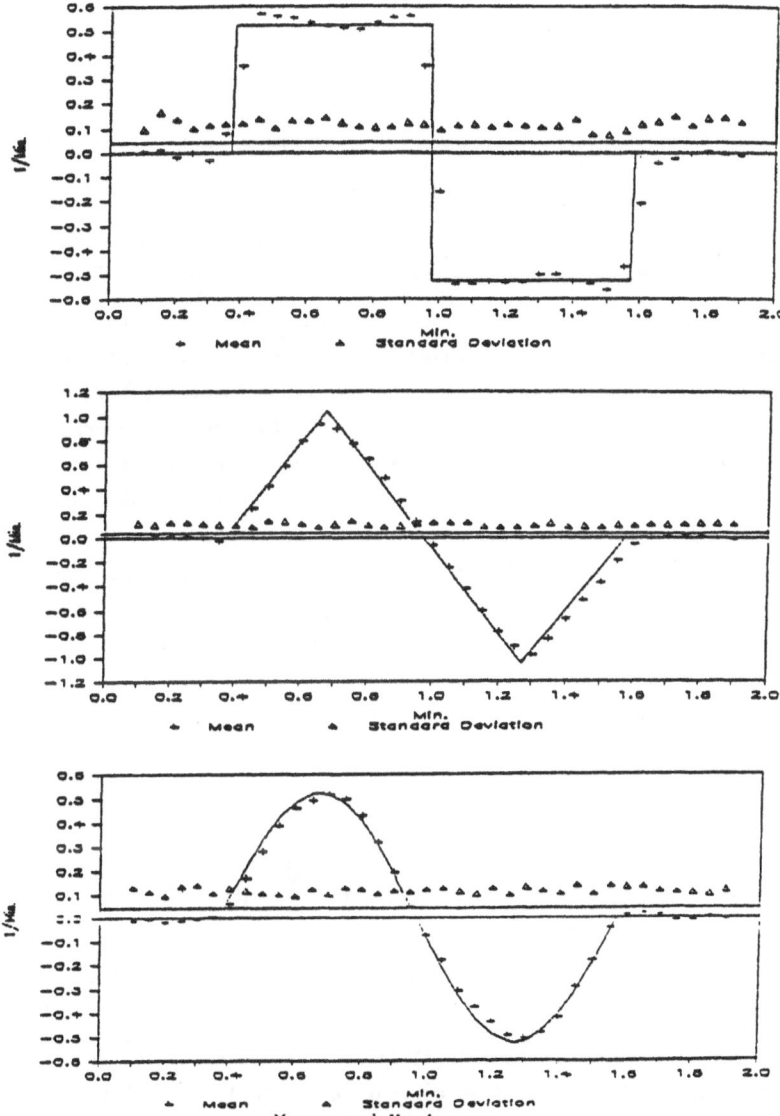

Figure 1. Estimation of three types of ·angular-velocity.

Hence,

$$
\hat{\omega} = \frac{\displaystyle\sum_{k=-n}^{n} (z_{2,k+2}-z_{2,k+1})^T B (z_{2,k+1}-z_{2,k})}{\displaystyle\sum_{k=-n}^{n} (z_{2,k+1}-z_{2,k})^T B^T B (z_{2,k+1}-z_{2,k})}
$$

$$
= \frac{\displaystyle\sum_{k=-n}^{n} (y_{k+2}-y_{k+1})(x_{k+1}-x_k)-(x_{k+2}-x_{k+1})(y_{k+1}-y_k)}{\delta t \displaystyle\sum_{k=-n}^{n} \left\{ (x_{k+1}-x_k)^2+(y_{k+1}-y_k)^2 \right\}} . \tag{12}
$$

When $\omega = 0$, $\sigma^2 = 0$ and there is no measurement error, the point object moves along a a straight line, $x_{k+1}-x_{k+1} = x_{k+1}-x_k$, $y_{k+2}-y_{k+1} = y_{k+1}-y_k$, so $\hat{\omega} = 0$ as it should.

Three types of angular-velocity variations are simulated, viz.: (1) a constant positive angular velocity followed by a constant negative one, (2) a triangular type, and (3) a sinusoidal type. The magnitude in (1), (3) is $\pi/6$ 1/min. and in (2) is $\pi/3$ 1/min. The total time interval is 2 min., and maneuvering begins at 0.38 min. and ends at 1.58 min. In all three types, the intensity of the Wiener process is 0.04^2 and measurement period is 0.01 min. The number of the summation is chosen as $2n+1=11$. The estimations of angular-velocity are shown in Figure 1. Here, 25 sample runs are used to obtain the statistics of estimation. The mean is quite close to the original maneuver in both magnitude and shape, while the standard deviation is almost 2 to 3 times that of the original one.

3. ARMA-TYPE MODELS

An affine bilinear, autoregressive, moving-average model may be defined by

$$
x_k = \sum_{i=1}^{p} a_i x_{k-i} + \sum_{i=1}^{p}\sum_{j=1}^{q} b_{j,i} x_{k-i} u_{k-j} + \sum_{j=0}^{q} c_j u_{k-j} + d . \tag{13}
$$

This is equivalent to a serial connection of several ABTS, i.e. model (13) defined as $AB(p,z)$. When $d = 0$ and $b_{i,j} = 0$, $\forall\ i,j$ then (13) becomes an $ARMA(p,q)$.

The generalized model $AB(p,q)$ is not readily amenable for transformation to vector form $AB(1,1)$,--found so convenient for $ARMA(p,q)$. But a special case of $AB(p,p)$, i.e., the non-anticipative $AB(p,p)$,

$$
x_k = \sum_{i=1}^{p} a_i x_{k-i} + \sum_{j=1}^{p}\sum_{i\geq j}^{p} b_{j,i} x_{k-i} u_{k-j} + \sum_{j=0}^{p} c_j u_{k-j} + d , \tag{14}
$$

can be transformed to a $2p-1$ dimensional vector from $AB(1,1)$ such that

$$
\begin{pmatrix} x \\ y_2 \\ \cdot \\ y_p \\ \overline{} \\ z_2 \\ z_3 \\ \cdot \\ z_p \end{pmatrix}_k - \begin{pmatrix} a_1 & & & & \\ a_2 & I_{p-1} & & \cdot & 0 \\ \cdot & & & & \\ a_p & 0 & \cdots & 0 & \\ 1 & & & 0 \cdots 0 \\ 0 & & & & \\ \cdot & & 0 & I_{p-2} & \cdot \\ 0 & & & & 0 \end{pmatrix} \cdot \begin{pmatrix} x \\ y_2 \\ \cdot \\ y_p \\ \overline{} \\ z_2 \\ z_3 \\ \cdot \\ z_p \end{pmatrix}_{k-1} + \begin{pmatrix} c_0 \\ 0 \\ \cdot \\ 0 \\ \overline{} \\ 0 \end{pmatrix} u_k + \begin{pmatrix} c_1 \\ c_2 \\ \cdot \\ c_p \\ \overline{} \\ 0 \end{pmatrix} u_{k-1} + \begin{pmatrix} d \\ 0 \\ \cdot \\ 0 \\ \overline{} \\ 0 \end{pmatrix}
$$

$$
+ \begin{pmatrix} b_{11} & b_{12} & b_{13} & \cdot & b_{1p} \\ b_{22} & b_{23} & b_{24} & b_{2p} & 0 \\ \cdot & 0 & \cdot & & 0 \\ b_{pp} & 0 & & & 0 \\ & & & & \\ 0 & & \cdot & & 0 \end{pmatrix} \begin{pmatrix} x \\ y_2 \\ \cdot \\ y_p \\ \overline{} \\ z_2 \\ z_3 \\ \cdot \\ z_p \end{pmatrix}_{k-1} \cdot u_{k-1} \, , \tag{15}
$$

i.e. $x_k = A x_{k-1} + B x_{k-1} u_{k-1} + C_0 u_k + C_1 u_{k-1} + D$,

$\quad x_k = H^T x_k$,

where $H^T = (1,0,\ldots,0)$. The proof is one of algebraic substitution.

Like the continuous bilinear stochastic process, the AB time series has a useful property, which makes analysis and identification convenient. For the AB(p,q) time series

$$
x_k = \sum_{i=1}^{p} a_i x_{k-i} + \sum_{i=1}^{p} \sum_{j=i}^{q} b_{i,j} x_{k-i} v_{k-j} + \sum_{j=0}^{p} c_j v_{k-j} + d \tag{17}
$$

or non-anticipative AB(p,p):

$$
x_k = A x_{k-1} + B x_{k-1} v_{k-1} + C_0 v_k + C_1 v_{k-1} + D \, ,
$$

$$
x_k = H^T x_k \, , \tag{18}
$$

where u_k, x_k are scalar input and output,

$\quad x_k$ is a $2p+1$ dimensional state vector,

$\quad A$, B, C_0, C_1, D are appropriately dimensional matrices,

$\quad H^T = [1,0,\ldots,0]$.

(1) If the input v_k is a static affine linear transform of u_k

$$v_k = m_v + \sigma_v u_k ,$$

where u_k is a standard white Gaussian random sequence, then the output x_k and state x_t are still of the form (17) or (18), i.e.,

$$x_k = \sum_{i=1}^{p} \left(a_i + m_v \sum_{j=1}^{1} b_{i,j} \right) x_{k-i} + \sigma_v \sum_{i=1}^{p} \sum_{j=1}^{q} b_{i,j} x_{k-i} v_{k-j}$$

$$+ \sigma_v \sum_{j=0}^{q} c_j v_{k-j} + d + m_x \sum_{j=0}^{q} c_j \tag{19}$$

or

$$\mathbf{x}_k = (A + Bm_v)\mathbf{x}_{k-1} + (B\sigma_v)\mathbf{x}_{k-1}u_{k-1} + (C_0\sigma_v)u_k + (C_1\sigma_v)u_{k-1} + (d + C_0 m_v + C_1 m_v) ,$$

$$x_k = H^T \mathbf{x}_k . \tag{20}$$

(2) If the output x_k is $x_k = m_x + \sigma_x y_k$, then the new output y_k and the new state $\mathbf{y}_k$ are still of the form (7) or (8)

$$y_k = \sum_{i=1}^{p} a_i y_{k-i} + \sum_{i=1}^{p} \sum_{j=1}^{q} b_{i,j} x_{k-i} v_{k-j} + (c_0/\sigma_x) v_k$$

$$+ \sum_{j=1}^{q} \left(c_j + m_x \sum_{i=1}^{p} b_{i,j} \right) v_{k-j}/\sigma_x + \left(d + m_x \left(\sum_{i=1}^{p} a_i - 1 \right) \right) \Bigg/ \sigma_x \tag{21}$$

or

$$\mathbf{y}_k = A\mathbf{y}_{k-1} + B\mathbf{y}_{k-1}v_{k-1} + C_0 v_k + C_1 v_{k-1} + Dm_x ,$$

$$y_k = H^T \mathbf{y}_k , \tag{22}$$

where

$$C_0^T = (c_0', 0, \ldots, 0) ,$$

$$C_j^T = (c_1', c_2', \ldots, c'_p, 0, \ldots, 0) ,$$

$$D^T = (d', 0, \ldots, 0)$$

$$c_0' = c_0/\sigma_x ,$$

$$c_i' = \left(c_i + m_x \sum_{j=i}^{p} b_{i,j} \right)/\sigma_x , \quad i=1, \ldots, p$$

$$d' = \left(d + \left(-1 + \sum_{i=1}^{p} a_i \right) m_x \right)/\sigma_x .$$

4. STATIONARITY AND MOMENT ESTIMATION

In this section the non-anticipative AB(p,p) (16), is studied, where the initial state x_0 is independent of u_1, $u_2,\ldots$.

It is obvious that: (i) x_k, x_k is independent of u_{k+1}, $u_{k+2},\ldots$; (ii) x_k and x_k are non-Gaussian, so that the first and second moments alone do not characterize the time series. To describe the time series all of the moments are required in general.

<u>Definition</u>. For the time series x_k, if the ℓth-order moment $E[\&\{x_k^{\ell}\}]$ is constant for any k then we say x_k is ℓth-order stationary; if the ℓth-order moment tends to a constant as k tends to ∞, then x_k is ℓth-order asymptotically stationary.

Here, $\&\{\cdot\}$ denotes summation over all possible product combinations (Kronecker products) such as discussed in the Appendix.

<u>Theorem</u>. For non-anticipative AB(p,p), if all the absolute values of the eigenvalues of the first ℓ following matrices

$$\sum_{i=0}^{[n/2]} \&\{A^{n-2i}B^{2i}\}(2i)!/2^i i! \qquad n=1,2,3,\ldots,1 \tag{23}$$

are less than 1, where $[n/2]$ is the largest integer no larger than n/2, then x_k, x_k are ℓth-order asymptotically stationary. The proof is not given here for brevity, but will be published elsewhere.

In analyzing an affine bilinear time series usually only one sample series is available and time-averages are used to estimate ensemble averages, just as for the linear time series case. Consequently, the following may be proven.

<u>Theorem</u>. For non-anticipative AB(p,p), if x_k is ℓth-order asymptotically stationary, then ℓth-order time-averaged correlated moments tend to an ensemble average; i.e., for any integer, $k_1,k_2,\ldots,k_\ell$,

$$(1)\ \lim_{N\to\infty} E\left[(1/N) \sum_{k=1}^{N} x_k \otimes x_{k+k_2} \otimes x_{k+k_3} \otimes \cdots \otimes x_{k+k_\ell} \right]$$
$$- \lim_{k\to\infty} E[x_k \otimes x_{k+k_2} \otimes x_{k+k_3} \otimes \cdots \otimes x_{k+k_\ell}] < \infty , \tag{24}$$

$$(2)\ \lim_{N\to\infty} E\left[(1/N) \sum_{k=1}^{N} x_{k+k_1} \otimes x_{k+k_1+k_2} \otimes x_{k+k_1+k_3} \otimes \cdots \otimes x_{k+k_\ell+k_\ell} \right]$$
$$- \lim_{N\to\infty} E\left[(1/N) \sum_{k=1}^{N} x_k \otimes x_{k+k_2} \otimes x_{k+k_3} \otimes \cdots \otimes x_{k+k_\ell} \right] ; \tag{25}$$

⊗ denotes Kronecker product, and is defined in the Appendix. Again, the proof will be given elsewhere.

5. SCALAR MODEL ESTIMATION

For the linear time series, which is Gaussian, only the first and second-order moments are needed to estimate parameters. For the bilinear time series, however, these are not enough. It can be shown for AB(1,1) that

$$E[x_{k+j+1}x_k] = aE[x_{k+j}x_k] + (bc_0 + d)E[x_k] , \qquad j \geq 1 , \qquad (26)$$

$$E[x_{k+j+2}x_{k+j+1}x_k] = (a^2 + b^2) E[x_{k+j+1}x_{k+j}x_k]$$

$$+ f(a,b,c_0,c_1,d)E[x_{k+j}x_k] + g(a,b,c_0,c_1,d)E[x_k] , \quad j \geq 1 , \qquad (27)$$

where

$$f = (a^2 - b^2)d + ad + 2abc_1 + 2c_0(a^2b + ab - b^3) ,$$

$$g = d^2 + a(d^2 + c_1^2) + c_0\Big((a^2 - b^2 + 1)c_1 + 2abd + 3bd \Big) + c_0^2\Big(a(1 + 2b^2) + 2b^2\Big) .$$

Here a, b, c_0, c_1, d replace A, B, C_0, C_1, D, and x_k replaces x_k, in (1) to designate scalars.

If only the first and second-order moments are known, from (26) we only can solve a and $bc_0 + d$. However, if at least the first three moments are available, it might be possible to solve parameters b, c_0, c_1, and d. Using time-average moment estimates for sample computations, it is convenient to define the following:

$$s \qquad = (1/N) \sum_{k=1}^{N} x_k , \qquad (28)$$

$$s_j \qquad = (1/N) \sum_{k=1}^{N} x_{k+j}x_k , \qquad (29)$$

$$s_{j+1,j} = (1/N) \sum_{k=1}^{N} x_{j+1+k}x_{j+k}x_k , \qquad (30)$$

and

$$\alpha_j = s_{j+1} - as_j - (bc_0 + d) s , \qquad (31)$$

$$\beta_j = s_{j+2,j+1} - (a^2 + b^2) s_{j+1,j} - fs_j - gs . \qquad (32)$$

If x_k is 6th-order asymptotically stationary, it can be shown that

$$E[\alpha_j] = 0 , \tag{33}$$

$$E[\beta_j] = 0 ; \tag{34}$$

$$\lim_{N \to \infty} E[\alpha_j^2] = 0 ; \tag{35}$$

$$\lim_{N \to \infty} E[\beta_j^2] = 0 . \tag{36}$$

For a given sample of x_k and sufficiently large N, we can calculate s, s_j, $s_{j+1,j}$ for $j=1,2,\ldots,L$. Then the estimators a, b, c_0, c_1, d are selected so as to minimize

$$\sum_{j=1}^{L-1} \alpha_j^2 \quad \text{and} \quad \sum_{j=1}^{L-1} \beta_j^2 .$$

For simplicity, a and $(bc_0 + d)$ could be estimated by minimizing $\sum \alpha_j^2$, i.e.

$$\hat{a} = \frac{L \sum_{j=1}^{L-1} s_{j+1} s_j - \sum_{j=1}^{L-1} s_{j+1} \sum_{j=1}^{L-1} s_j}{L \sum_{j=1}^{L-1} s_j^2 - \left(\sum_{j=1}^{L-1} s_j^2 \right)^2} , \tag{37}$$

$$b\hat{c}_0 + d = \frac{\sum_{j=1}^{L-1} s_j^2 \sum_{j=1}^{L-1} s_{j+1} - \sum_{j=1}^{L-1} s_j \sum_{j=1}^{L-1} s_j s_{j+1}}{sL \sum_{j=1}^{L-1} s_j^2 - \left(\sum_{j=1}^{L-1} s_j^2 \right)^2} , \tag{38}$$

$a^2 + b^2$, f, g are estimated by minimizing β_j^2. Then estimate b, c_0, c_1, d, or use nonlinear programming directly.

When $c_0 = 0$, x_k is independent of u_k, $u_{k+1},\ldots$, the computation becomes much easier. Now have

$$E[x_{k+j+1}x_k] = aE[x_{k+j}x_k] + dE[x_k] , \qquad j \geq 0 , \tag{39}$$

$$E[x_{k+j+2}x_{k+j+1}x_k] = (a^2 + b^2) E[x_{k+j+1}x_{k+j}x_k]$$
$$+ \left((a^2 - b^2) d + ad + 2abc_1 \right) E[x_{k+j}x_k]$$
$$+ \left(d^2 + a(d^2 + c_1^2) \right) E[x_k] , \qquad j \geq 0 , \tag{40}$$

$$E[x_{k+j+1}^2 x_k] = (a^2 + b^2) E[x_{k+j}^2 x_k] + 2(bc_1 + ad) E[x_{k+j}x_k]$$
$$+ (c_1^2 + d^2) E[x_k] , \qquad j \geq 0 . \tag{41}$$

Let s, s_j be the same as (28), (29) and

$$s_{j,j} - (1/N) \sum_{k=1}^{N} x_{k+j}^2 x_k , \quad j \geq 0 , \tag{42}$$

$$\alpha_j - s_{j+1} - a \cdot s_j - d \cdot s , \tag{43}$$

$$\beta_j - s_{j+1,j+1} - (a + b^2) s_{j,j} - 2(bc_1 + ad) s_j - (c_1^2 + d^2) s . \tag{44}$$

Then, it can be proven that

$$E[\alpha_j] - E[\beta_j] - 0 , \qquad \lim_{N \to \infty} E[\alpha_j^2] - \lim_{N \to \infty} E[\beta_j^2] - 0 .$$

From minimizing $\sum_{j=0}^{L-1} \beta_j^2$, $L \geq 2$,

$$\hat{a} - \frac{L \sum_{j=0}^{L-1} s_{j+1} s_j - \sum_{j=0}^{L-1} s_{j+1} \sum_{j=0}^{L-1} s_j}{L \sum_{j=0}^{L-1} s_j^2 - \left(\sum_{j=0}^{L-1} s_j^2 \right)^2} , \tag{45}$$

$$\hat{d} - \frac{\sum_{j=0}^{L-1} s_j^2 \sum_{j=0}^{L-1} s_{j+1} - \sum_{j=0}^{L-1} s_j \sum_{j=0}^{L-1} s_j s_{j+1}}{sL \sum_{j=0}^{L-1} s_j^2 - \left(\sum_{j=0}^{L-1} s_j^2 \right)^2} . \tag{46}$$

Then substitution in (44), and minimizing $\sum_{j=0}^{L-1} b_j^2$ yields estimates of b and

c_1. Here notice that $\hat{b}$, $\hat{c}_1$ and $-\hat{b}$, $-\hat{c}_1$ are two equal estimators.

Below a numerical simulation result is provided for a first-order bilinear times series:

$$x_k - 0.4x_{k-1} + 0.3x_{k-1}u_{k-1} + u_{k-1} .$$

After checking, x_k is 6-th order asymptotically stationary. For $x_0 - 0$, $n - 1000$, $L - 3$, a sample of x_k is generated and s, s_j, $s_{j,j}$ are calculated as follows:

	simulation value	asymptotical value
s	-0.0576	0.0
s_0	1.3150	1.3333
s_1	0.4853	0.5333
s_2	0.1833	0.2133
s_3	0.0721	0.0853
$s_{0,0}$	1.1548	1.159
$s_{1,1}$	0.9621	1.089
$s_{2,2}$	0.3764	0.592
$s_{3,3}$	0.0756	0.120

The estimators of parameters are

$$\hat{a} = 0.365, \quad \hat{d} = -0.0979, \quad \hat{b} = 0.329, \quad \hat{c}_1 = 0.909 .$$

where $\hat{b}$, $\hat{c}_1$ are yielded by a gradient method.

The moments method is generalized from correlation methods for linear time series. It can be useful, especially when a lot of data are available over the time interval.

ACKNOWLEDGEMENT

This research was supported by ONR Contract N00014-81-K-0814.

REFERENCES

1. V.I. Tatarskie, *Wave Propagation in a Turbulent Medium*, McGraw-Hill, New York, 1961.

2. R.R. Mohler, W.J. Kolodziej, R.S. Engelbrecht, and H.D. Brunk, "On Nonlinear Tracking and Filtering," in *Statistical Signal Processing* (E.Wegman, J. Smith, eds.), Marcel Dekker, New York, 1984.

3. R.R. Mohler, *Bilinear Control Processes*, Academic Press, New York, 1973.

4. P.A. Cartwright, "A Note on Using State Dependent Models with a Time Dependent Variance," *J. Business and Economic Statistics 2*, 410-413, 1984.

5. P.A. Cartwright, "Forecasting Time Series: A Comparative Analysis of Alternative Classes of Time Series Models," *J. Time Series Analysis 6*, 203-211, 1985.

6. V. Haggan and T. Ozaki, "Amplitude Dependent AR Model Fitting for Nonlinear Random Vibrations," *Int'l Time-Series Meeting, Nottingham*, March 1979.

7. M.B. Priestley, "State-Dependent Models: A General Approach to Nonlinear Time Series Analysis," *J. Time Series Analysis 1*, 47-71, 1980.

8. M.B. Priestley, "On the Fitting of General Nonlinear Time Series Models," in *Time Series Analysis: Theory and Practice I* (D.D. Anderson, ed.), North-Holland, Amsterdam, 717-731, 1981.

9. M.B. Priestley, *Probability and Mathematical Statistics: Spectral Analysis and Series*, Academic Press, New York, 1982.

10. M.B. Priestley, "A Study of the Application of State-Dependent Models in Nonlinear Time Series Analysis," *J. Time Series Analysis 3*, 69-102, 1984.

11. T.S. Rao, "On the Theory of Bilinear Time SEries Models," *J.R. Statistics Soc. B40*, 244-255, 1981.

12. M.M. Gabr and T.S. Rao, "The Estimation and Prediction of Subset Bilinear Time Series Models with Applications," *J. Time Series Analysis 2*, 155-171, 1981.

13. D.T. Pham and L.T. Tran, "On the First-Order Bilinear Time Series Model," *J. Appl. Prob. 18*, 617-627, 1981.

14. H. Tong, "A Note on a Markov Bilinear Stochastic Process in Discrete Time," *J. Time Series Analysis 2*, 279-285, 1981.

15. B.W. Shou-Ren and A. Hong-Zhi, "On the Distribution of a Simple Stationary Bilinear Process," *J. Time Series Analysis 4*, 209-216, 1983.

16. M.B. Rao, T.S. Rao, and A.M. Walker, "On the Existence of Some Bilinear Time Series Models," *J. Time Series Analysis 4*, 95-110, 1983.

17. K. Kumar, "On the Identification of Some Bilinear Time Series Models," *J. Time Series Analysis 7*, 117, 1986.

18. E.M. Engle, "A Unified Approach to the Study of Sums, Products, Time-Aggregation and Other Functions of ARMA Processes," *J. Time Series Analysis 5*, 159, 1984.

19. P.D. Feigin and R.L. Tweedie, "Random Coefficient Autoregressive Processes," *J. Time Series Analysis 6*, 1, 1985.

20. W.K. Li, "On the Autocorrelation Structure and Identification of Some Bilinear Time Series," *J. Time Series Analysis 5*, 173-181, 1984.

21. D.F. Nicholls and B.G. Quinn, *Random Coefficient Autoregressive Models: An Introduction*, Springer-Verlag, Lecture Notes in Statistics, New York, 1982.

22. C.W. Granger and A.P. Anderson, *An Introduction to Bilinear Time Series Models*, Vandenhoech and Ruprecht, Gotingen, 1978.

23. M.J. Hinich, "Testing for Gaussianity and Linearity of a Stationary Time Series," *J. Time Series Analysis 3*, 169-176, 1982.

24. T.S. Rao and M.M. Gabr, "A Test for Linearity of Stationary Time Series," *J. Time Series Analysis 1*, 145-158, 1980.

25. R.R. Mohler, C. Bruni, and A. Gandolfi, "A Systems Approach to Immunology," *Proc. IEEE 68*, 964-990, 1980.

26. R.R. Mohler, "Foundations of Immune Control and Cancer," *Recent Advances in System Science* (A.V. Balakrishnan, ed.), Optimization Software Publication, New York, to appear.

27. W.J. Kolodziej and R.R. Mohler, "State Estimation of Conditionally Linear Systems," *SIAM J. Cont. & Optimiz. 24*, 497-508, 1986.

APPENDIX: KRONECKER ALGEBRA

Definition

Suppose A is a mxn matrix, B is a kxℓ matrix, the Kronecker product of A, B is a (mk)x(nℓ) matrix.

$$A \otimes B = \begin{pmatrix} a_{11}B & a_{12}B & \cdots & a_{1n}B \\ \cdot & \cdot & \cdots & \cdot \\ a_{m1}B & a_{m2}B & \cdots & a_{mn}B \end{pmatrix}$$

Properties

Suppose A, B, C, D are appropriate dimensional matrices; I is an appropriate dimensional unity matrix; then

(1) although $A \otimes B$ and $B \otimes A$ have the same dimension, generally $A \otimes B \neq B \otimes A$;

(2) $A \otimes 1 = 1 \otimes A = A$, (1 is a scalar)

(3) $(A+B) \otimes C = A \otimes C + B \otimes C$,

$\qquad A \otimes (B+C) = A \otimes B + A \otimes C$;

(4) $(AB) \otimes (CD) = (A \otimes C)(B \otimes D)$,

$\qquad$ if C is an mxn matrix, then

$$(AB) \otimes C = (A \otimes I_m)(B \otimes C) = (A \otimes C)(B \otimes I_n);$$

(5) if A, B are proper matrices, then

$$(A \otimes B)^{-1} = A^{-1} \otimes B^{-1} .$$

Notation

Suppose there exit i A matrices and j B matrices; the Kronecker products of any combinations of A's and B's have the same dimension. Let $\&(A^i B^j)$ denote the summation of all possible Kronecker product combinations of i A's and j B's. For example

$$\&(A^2) \quad = A \otimes A ,$$

$$\&(AB) \quad = A \otimes B + B \otimes A ,$$

$$\&(AB^2) \quad = A \otimes B \otimes B + B \otimes A \otimes B + B \otimes B \otimes A ,$$

$$\&(A^2 B^2) = A \otimes A \otimes B \otimes B + A \otimes B \otimes A \otimes B + A \otimes B \otimes B \otimes A + B \otimes A \otimes B \otimes A + B \otimes A \otimes A \otimes B + B \otimes B \otimes A \otimes A .$$

Eigenvalues of Kronecker Product

(1) Suppose $(r_{a1}, \ldots, r_{an})$, $(r_{b1}, \ldots, r_{bn})$ are respective eigenvalues of nxn matrices A and B; then $(r_{ai} r_{bj}, i,j = 1, \ldots, n)$ are the eigenvalues of $A \otimes B$. In fact, there exist proper matrices P and Q such that

$$A = P \Lambda(r_{ai}) P^{-1} , \qquad B = Q \Lambda(r_{bj}) Q^{-1} ,$$

where Λ is a diagonal matrix.

Now

$$A \otimes B = (P \otimes Q)(\Lambda(r_{ai}) \otimes \Lambda(r_{bj}))(P^{-1} \otimes Q^{-1})$$

$$= (P \otimes Q)(\Lambda(r_{ai}r_{bj}))(P \otimes Q)^{-1}.$$

(2) The eigenvalues of $(A \otimes I_n)$ consist of n duplicate r_{ai}, $i=1,\ldots,n$.

(3) The eigenvalues of $\&(AI_n)$ are $\{r_{ai}+r_{aj}, i,j = 1,\ldots,n\}$. In fact

$$\&(AI_n) = A \otimes I_n + I_n \otimes A$$

$$= P\Lambda(r_{ai})P^{-1} \otimes I_n + I_n \otimes P\Lambda(r_{aj})P^{-1}$$

$$= P(\Lambda(r_{ai}, i,j=1,\ldots,n))P^{-1} + P(\Lambda(r_{aj}, i,j=1,\ldots,n))P^{-1}$$

$$= P(\Lambda(r_{ai} + r_{aj}, i,j=1,\ldots,n))P^{-1} .$$

BIVARIATE BILINEAR MODELS AND THEIR SPECIFICATION

Kuldeep Kumar
Indian Institute of Management
Lucknow, India

1. INTRODUCTION

In many forecasting situations other events will systematically influence the series to be forecasted and we can use the information contained in other related time series. The models that incorporate more than one time series and introduce explicitly the dynamic characteristic of the system are called multiple time series models. In particular, the models having only two stationary time series X_t and Y_t are called bivariate time series models.

Most of the working bivariate time series models has been done considering the relationship between the dependent variable Y_t and the explanatory variable X_t as linear. But it is quite possible that for some y_t and x_t the relationship may not be linear, but it may follow some other type of nonlinear, or say bilinear, relationship. It has been found by Subba Rao and Gabr (1984) that some time series like monthly unemployment figures in West Germany are nonlinear and the forecasts obtained by bilinear models are better than the forecasts obtained from linear models. However, it has been observed by them that a series like unemployment figures is influenced by many other variables and it would be interesting to study the multivariate extension of bilinear time series models. The aim here is to extend the already developed concept of bilinear models for univariate series to bivariate bilinear models. Granger and Andersen (1978, p. 89) considered briefly the possibility of extensions of bilinear models to bivariate bilinear models and have mentioned that "more extreme" examples of the deficiencies of covariance techniques occur when bivariate bilinear models are considered. Subba Rao (1986) has defined a multivariate extension of the bilinear time series models defined earlier for the univariate case by Subba Rao (1981) and studied some of the statistical properties of these processes in a particular case.

In this chapter we have defined bivariate bilinear models which could be considered as a special case of the multivariate extension of the bilinear time series models. We are more concerned here with the specification problem of these bivariate bilinear models and to achieve this we have extended the concept of cross correlation functions to third order cross moments. It may be mentioned that Kumar (1986a) has used the similar concept of third order moments to specify some of the bilinear time series models.

In Section 2 we have given important definitions and in Section 3 we have given some theoretical results which are useful in the identification of bivariate bilinear models. In Section 4 we have carried out some simulation studies to demonstrate the identification procedure and to verify the theoretical results given in Section 3. Finally, conclusions have been drawn in Section 5.

2. DEFINITIONS

Definition 1: The general bivariate bilinear model could be defined as follows:

$$y_t = \sum_{k'=1}^{q} \beta_{k'} y_{t-k'} + \sum_{l'=1}^{P} \alpha_{l'} x_{t-l'} + \sum_{k'=1}^{Q} \sum_{l'=0}^{P} \beta_{k'l'} y_{t-k'} x_{t-l'} + a_t \tag{2.1}$$

where x_t is the input series which could be considered as white noise either in its original form or after prewhitening. So

$$E(x_t) = 0$$
$$E(x_t x_{t+k}) = \sigma_x^2 \quad \text{if} \quad k = 0$$
$$= 0 \quad \text{if} \quad k \neq 0$$

The a_t's are independently, identically and normally distributed random variables with mean zero and variance σ_a^2.

Definition 2: The completely bilinear model could be defined as

$$y_t = \sum_{k'=1}^{Q} \sum_{l'=1}^{P} \beta_{k'l'} y_{t-k'} x_{t-l'} + a_t \tag{2.2}$$

There are a number of special cases of the completely bivariate bilinear models as mentioned in the following definitions.

Definition 3: A completely bivariate bilinear subdiagonal model can be written in the general form as

$$y_t = \sum_{\substack{k'=1 \\ k'<l'}}^{Q} \sum_{l'=2}^{P} \beta_{k'l'} y_{t-k'} x_{t-l'} + a_t$$

and an example is

$$y_t = \beta y_{t-1} x_{t-2} + a_t \tag{2.3}$$

Definition 4: A completely bivariate bilinear diagonal model can be defined as

$$y_t = \sum_{k'=0}^{P} \beta_{k'} y_{t-k'} x_{t-k'} + a_t$$

and an example of this model is

$$y_t = \beta y_{t-1} x_{t-1} + a_t \tag{2.4}$$

Definition 5: A completely bivariate bilinear superdiagonal model can be defined as

$$y_t = \sum_{\substack{k'=2 \\ k'>l'}}^{Q} \sum_{l'=1}^{P} \beta_{k'l'} y_{t-k'} x_{t-l'} + a_t$$

and an example of this model is

$$y_t = \beta y_{t-2} x_{t-1} + a_t \tag{2.5}$$

Definition 6: The cross covariance between x and y at lag(k) is defined as

$$\gamma_{xy}(k) = E[(x_t - \mu_x)(y_{t+k} - \mu_y)] \qquad k=0, \pm 1, \pm 2, \ldots$$

and the cross covariance between y and x at lag(k) is defined as

$$\gamma_{yx}(k) = E[(y_t - \mu_y)(x_{t+k} - \mu_x)]$$

In general, $\gamma_{xy}(k)$ will not be the same as $\gamma_{yx}(k)$, but $\gamma_{xy}(k) = \gamma_{yx}(-k)$.

The cross correlation function between x and y is defined as

$$\rho_{xy}(k) = \frac{\gamma_{xy}(k)}{\sigma_x \sigma_y}$$

The cross correlation measures the degree of association between the explanatory variable and the dependent variable at various lags.

Definition 7: The *third order cross moments* between variables x and y at lag (k,l) can be defined as

and the third order cross moments between variable y and x at lag (k,l) can be defined as

$$\gamma_{yx}(k,l) = E[(y_t-\mu_y)(x_{t+k}-\mu_x)(x_{t+l}-\mu_x)] \quad (k,l=0, \pm1, \pm2, \pm3,...) \tag{2.7}$$

In general $\gamma_{xy}(k,l)$ will not be equal to $\gamma_{yx}(k,l)$ and there will be no simple relationship between the two sets of cross moments.

Definition 8: If we arrange the third order cross moments, say $\gamma_{xy}(k,l)$ in a tab-ular form for different values of k and l, we get a third order cross moment table as shown below in Table 1. It may be mentioned

$$\gamma_{xy}(k,l) = \gamma_{xy}(l,k)$$

and so the elements in this table are symmetric.

Table 1. Third Order Cross Moment Table Between Variables x and y.

l \ k	. . .	-2	-1	0	1	2	. . .
.							
.							
.							
-2	. . .	$\gamma_{xy}(-2,-2)$	$\gamma_{xy}(-1,-2)$	$\gamma_{xy}(0,-2)$	$\gamma_{xy}(1,-2)$	$\gamma_{xy}(2,-2)$	. . .
-1	. . .	$\gamma_{xy}(-2,-1)$	$\gamma_{xy}(-1,-1)$	$\gamma_{xy}(0,-1)$	$\gamma_{xy}(1,-1)$	$\gamma_{xy}(2,-1)$	. . .
0	. . .	$\gamma_{xy}(-2,0)$	$\gamma_{xy}(-1,0)$	$\gamma_{xy}(0,0)$	$\gamma_{xy}(1,0)$	$\gamma_{xy}(2,0)$	. . .
1	. . .	$\gamma_{xy}(-2,1)$	$\gamma_{xy}(-1,1)$	$\gamma_{xy}(0,1)$	$\gamma_{xy}(1,1)$	$\gamma_{xy}(2,1)$	. . .
2	. . .	$\gamma_{xy}(-2,2)$	$\gamma_{xy}(-1,2)$	$\gamma_{xy}(0,2)$	$\gamma_{xy}(1,2)$	$\gamma_{xy}(2,2)$	. . .

3. SOME THEORETICAL RESULTS

In this section we have derived the third order cross moments for some simple bivariate bilinear models. We have assumed throughout our discussion that the input series x_t is prewhitened and the error series a_t is independently and identically distributed with mean zero and variance σ_a^2.

3.1 Bivariate Bilinear Diagonal Model

Let us take a single bivariate bilinear diagonal model (2.4)

$$y_t = \beta y_{t-1} x_{t-1} + a_t$$

Now

$$E(y_t) = \beta E(y_{t-1} x_{t-1}) = \beta E(\beta y_{t-2} y_{t-2} + a_{t-1}) x_{t-1}$$

$$= \beta^2 E[y*Xt-2x_{t-2} x_{t-1}] + E[a_{t-1} x_{t-1}]$$

$$= 0$$

as $\{x_t\}$ is white noise and $\{a_t\}$, $\{x_t\}$ are independent. Also,

$$\gamma_{xy}(k) = E[x_t y_{t+k}]$$

$$= E[x_t (\beta y_{t+k-1} x_{t+k-1} + a_{t+k})]$$

$$= \beta E[x_t x_{t+k-1} y_{t+k-1}]$$

$$= \beta E[x_t x_{t+k-1} (\beta y_{t+k-2} x_{t+k-2} + a_{t+k-1})]$$

$$= \beta^2 E[x_t x_{t+k-1} x_{t+k-2} y_{t+k-2}]$$

$$\vdots$$

$$= \beta^{k-1} E[x_t x_{t+k-1} x_{t+k-2} \cdots x_t y_{t-1} y_{t-1}]$$

$$= \beta^{k-1} E[x_t x_{t+k-1} \cdots x_t] E[x_{t-1} y_{t-1}]$$

$$= 0$$

Similarly, we can show

$$\gamma_{yx}(k) = 0$$

This shows that for the diagonal bivariate bilinear models all second order cross covariances are zero.

Now let us evaluate the third order cross moments

$$\gamma_{xy}(k,1) = E[x_t y_{t+k} y_{t+1}]$$

for k=1, 1=0

$$\gamma_{xy}(1,0) = E[x_t y_t y_{t+1}]$$

$$= E[(\beta y_t x_t + a_{t+1}) x_t y_t]$$

$$= \beta E[x_t^2 y_t^2]$$

$$= \beta E[y_t^2] E[x_t^2] = \beta E[y_t^2] \sigma_x^2 \qquad (3.1)$$

Now,

$$E[y_t^2] = E[\beta y_{t-1} x_{t-1} + a_t)]^2$$

$$= E[\beta^2 y_{t-1}^2 x_{t-1}^2 + a_t^2] + \text{cross product}$$

$$= \beta^2 E[y_{t-1}^2 x_{t-1}^2] + E[a_t^2]$$

$$= \beta^2 E[y_t^2 x_t^2] + \sigma_a^2 \quad \text{from stationarity}$$

$$= \beta^2 \sigma_x^2 E[y_t^2] + \sigma_a^2$$

So

$$E[y_t^2] = \frac{\sigma_a^2}{1-\beta^2 \sigma_x^2}$$

Substituting the value of $E[y_t^2]$ in (3.1) we get

$$\gamma_{xy}(1,0) = \frac{\beta \sigma_a^2 \sigma_x^2}{1 - \beta^2 \sigma_x^2}$$

It can be shown easily that for all other values of k,1; $\gamma_{xy}(k,1) = 0$.

Let us consider the third order cross moment between y and x

$$\gamma_{yx}(k,1) = E[y_t x_{t+k} x_{t+1}]$$

$$= E[(\beta y_{t-1} x_{t-1} + a_t) x_{t+k} x_{t+1}]$$

$$= \beta E[y_{t-1} x_{t-1} x_{t+k} x_{t+1}]$$

Let us take (k=0, l=0)

$$\gamma_{yx}(0,0) = \beta E[y_{t-1}x_{t-1}]E(x_t^2) = 0$$

In general for k=l=0, ±1, ±2,...

$$\gamma_{yx}(k,l) = 0$$

for k≠l, say k=1, l=2

$$\gamma_{yx}(1,2) = \beta E(y_{t-1}x_{t-1}x_{t+1}x_{t+2})$$

$$= 0$$

Similarly it can be shown that for other values of k,l third order cross moment $\gamma_{yx}(k,l)=0$.

The results given in this subsection can be summarized in the following lemma.

Lemma 1: For a simple diagonal bivariate bilinear model

$$y_t = \beta y_{t-1}x_{t-i} + a_t$$

all cross covariances are zero. However, for this particular model, the third order cross moments

$$\gamma_{xy}(k,l) = E[x_t y_{t+k} y_{t+l}]$$

are not zero when k=1, l=0 (and by symmetry k=0, l=1) but for all other values of k,l

$$\gamma_{xy}(k,l) = 0$$

Also, $\gamma_{yx}(k,l) = 0$ for all values of k and l.

In general for the diagonal bivariate bilinear model

$$y_t = y_{t-k'}x_{t-l'} + a_t \qquad \text{when } k'=l'$$

$\gamma_{xy}(k,l)$ is not zero when k=k', l=0 and k=0, l=k' but for all other values of k,l $\gamma_{xy}(k,l) = 0$. Also $c_{yx}(k,l) = 0$ for all k,l.

3.2 Bivariate Bilinear Subdiagonal Model

Let us consider a simple bivariate subdiagonal model

$$y_t = \beta y_{t-1} x_{t-2} + a_t$$

Now, $E(y_t) = 0$. Cross covariances for the above model are

$$\gamma_{xy}(k) = E(x_t y_{t+k}) = 0 \text{ for all } k$$

and also

$$\gamma_{yx}(k) = 0 \text{ for all } k.$$

i.e., cross covariances for this model are all zero. Let us consider the third order cross moments of x and y

$$\gamma_{xy}(k,l) = E[x_t y_{t+k} y_{t+l}]$$

for (k=2, l=1) or (k=1, l=2) we have

$$\gamma_{xy}(2,1) = E[x_t y_{t+2} y_{t+1}]$$

$$= E[(\beta y_{t+1} x_t + a_{t+2}) y_{t+1} x_t]$$

$$= \beta E[y_{t+1}^2 x_t^2]$$

$$= \beta E[y_{t+1}^2] \sigma_x^2 = \frac{\beta \sigma_a^2}{1 - \beta^2 \sigma_x^2} \sigma_x^2$$

$$\left(\text{since it can be shown that } E(y_t^2) = \frac{\sigma_a^2}{1 - \beta^2 \sigma_x^2} \right)$$

It can be proved easily that for all other values of (k,l), $\gamma_{xy}(k,l) = 0$.

It can be shown again that the third order cross moments of y and x

$$\gamma_{yx}(k,l) = E[y_t x_{t+k} x_{t+l}] = 0 \text{ for all values of } k,l.$$

The above results for the subdiagonal bivariate bilinear model can be summarized in the following lemma.

Lemma 2: For a simple subdiagonal bivariate bilinear model

$$y_t = \beta y_{t-1} x_{t-2} + a_t$$

all cross covariances are zero. However, for this model third order cross moments

$$\gamma_{xy}(k,l) = E(x_t y_{t+k} y_{t+l})$$

are not zero when k=2, l=1 or k=1, l=2 but they are zero for all other values of k,l. The third order cross moments

$$\gamma_{yx}(k,l) = E(y_t x_{t+k} x_{t+l})$$

are zero for all values of k,l=0,±1,±2,... .

In general for the subdiagonal bivariate bilinear model

$$y_t = \beta y_{t-k'} \, x_{t-l'} + a_t \qquad (k' < l')$$

$\gamma_{xy}(k,l)$ will not be zero when (k=k', l=l') or (k=l', l=k') and for all other values of k,l $\gamma_{xy}(k,l) = 0$. Also $\gamma_{yx}(k,l) = 0$ for all values of k and l.

3.3 Superdiagonal Bivariate Bilinear Models

Let us consider a simple superdiagonal bivariate bilinear model

$$y_t = \beta y_{t-2} x_{t-1} + a_t$$

It can be shown that

$$E(y_t) = 0$$

and also $\gamma_{xy}(k) = 0$, $\gamma_{yx}(k) = 0$ for all k.

Let us consider the third order cross moments

$$\gamma_{xy}(k,l) = E[x_t y_{t+k} y_{t+l}]$$

for (k=-1, l=1) or (k=1, l=-1) we have

$$\gamma_{xy}(-1,1) = E[x_t y_{t-1} y_{t+1}]$$

$$= E[(\beta y_{t-1} x_t + a_{t+1}) x_t y_{t-1}]$$

$$= \beta E[x_t^2 y_{t-1}^2]$$

$$= \frac{\beta \sigma_x^2 \sigma_a^2}{1 - \beta^2 \sigma_x^2}$$

since

$$E(y_t^2) = \frac{\sigma_a^2}{1 - \beta^2 \sigma_x^2}$$

It can be shown that for all other values of k, l $\gamma_{xy}(k, l) = 0$. Also it can be shown that the third order cross moments $\gamma_{yx}(k, l)$ are zero for all values of k and l.

From Sections 3.1, 3.2 and 3.3 it is clear that cross covariances at all lags are zero for diagonal, subdiagonal and superdiagonal bivariate bilinear models. Some of the third order cross moments are not zero in all the three cases. In the case of subdiagonal bivariate bilinear models, the third order cross moments will not be zero in some of the cells when $k, l > 0$. In the case of diagonal bivariate bilinear models the third order cross moments will not be zero when $(k=k'>0, l=0)$ or $(l=l'>0$ and $k=0)$ for some $k'=l'$. Similarly in the case of superdiagonal bivariate bilinear models the third order cross moments will not be zero for some values when $k>0$, $l<0$ or $k<0$, $l>0$.

4. SIMULATION RESULTS

In this section we have carried out some simulation studies to verify the theoretical results stated in Section 3, and to demonstrate the identification procedure.

Simulation studies were carried out on the ICL2960 at the University of Kent and NAG (mark 9) subroutine G05CCF and G05DDF were used to generate normal noise $\{a_t\}$ with mean zero and standard deviation one. Nag subroutine G05CAF was also used.

Models (2.3), (2.4) and (2.5) were studied. The value of β was varied from 0.2 to 0.70. Series length was varied from 100 to 300 observations and each experiment was repeated 50 times. The mean and standard deviation of sample third order cross moments were recorded in each case. For each experiment the first 50 observations were discarded to get rid of any "starting up" effect. The results were found to be quite satisfactory and agreed well with the theoretical results. A selection of results are given in Tables 2 to 4. These tables, which give estimates of $\gamma_{xy}(k, l)$ and corresponding standard deviation (in brackets) from simulation experiments for $k; l=-3, -2, \ldots, 2, 3$ are now discussed.

Table 2 gives the mean of the sample estimates of $\gamma_{xy}(k, l)$ and corresponding standard errors (in brackets) for different values of k, l for the diagonal bivariate bilinear model (2.4) when b=0.70. The length of the series is 200 and the experiment is repeated 50 times. It can be seen that simulation results agree very well

Table 2. Third Order Cross Moments $\gamma_{xy}(k,l)$ for the Diagonal Bivariate Bilinear Model $y_t - \beta y_{t-1} x_{t-1} + a_t$ where $\beta - 0.70$, $n - 200$ and Numbers of Iteration (I) — 50.

l \ k	-3	-2	-1	0	1	2	3
-3	-0.01 (0.19)	-0.00 (0.14)	-0.01 (0.15)	-0.02 (0.13)	-0.01 (0.16)	-0.00 (0.16)	-0.00 (0.16)
-2	-0.00 (0.14)	-0.00 (0.18)	0.01 (0.17)	-0.01 (0.17)	-0.01 (0.21)	-0.00 (0.14)	0.01 (0.17)
-1	-0.01 (0.15)	0.01 (0.17)	-0.02 (0.21)	0.01 (0.21)	-0.02 (0.20)	0.04 (0.19)	0.00 (0.16)
0	-0.02 (0.13)	-0.01 (0.17)	0.01 (0.21)	-0.01 (0.20)	1.37 (0.27)	-0.02 (0.15)	0.00 (0.21)
1	-0.01 (0.16)	-0.01 (0.21)	-0.02 (0.20)	1.37 (0.27)	0.04 (0.17)	0.03 (0.16)	-0.04 (0.12)
2	-0.00 (0.16)	-0.00 (0.14)	0.04 (0.19)	-0.02 (0.15)	-0.03 (0.16)	-0.04 (0.18)	0.02 (0.18)
3	-0.00 (0.14)	0.01 (0.17)	0.00 (0.16)	0.00 (0.21)	-0.04 (0.22)	0.02 (0.18)	-0.04 (0.11)

Theoretical value in the cell $\gamma_{xy}(1,0) - \gamma_{xy}(0,1)$ is 1.42, and the rest of the theoretical values are zero.

with the theoretical results and the values in the cell (k-1, l-0 or k-0, l-1) are much larger than any other value. The value corresponding to this cell is 1.37 and the maximum of other values is only 0.04. Theoretically we would expect the values in the cell (k-1, l-0 or k-0, l-1) to be equal to 1.42 and the values in the other cell to be equal to zero. We have carried out some individual simulation studies and the other individual simulation results are more or less similar to those for univariate diagonal, bilinear time series models discussed in Kumar (1986b, Ch.3). It was found that for series of moderate length (say between 100-250 observations) and values of $\beta > 0.5$ the pattern was distinguishable but it may not be distinguishable when the length of the series is small or the values of β is quite low.

We have also obtained the third order cross moments $\gamma_{yx}(k,l)$ for different values of k and l for the same model (2.4) but it was found as expected theoretically that all estimates of $\gamma_{yx}(k,l)$ are quite close to zero.

Table 3 gives the mean of the sample estimate of the third order cross moments $\gamma_{xy}(k,l)$ and corresponding standard errors (in brackets) for the subdiagonal bivariate bilinear model (2.3) when $\beta-0.70$, the length of the series is 200 and the experiment was repeated 50 times. In this case also the simulation results agree quite closely with the theoretical results. As mentioned in Lemma 6.2, the values corresponding to (k-2,l-1 or k-1, l-2) are dominating over other values. The value corresponding to these cells is 1.35 while the maximum of other values is only 0.04. Theoretically the expected value in the cells (k-2, l-1 or k-1, l-2) is 1.42 and the rest of the entries are theoretically zero.

The estimates of the third order cross moments $\gamma_{yx}(k,l)$ for different values of (k,l) were found to be very close to zero as expected theoretically for the same model.

We have also studied the superdiagonal bivariate bilinear model (2.5) when $\beta-0.70$. Table 4 gives the mean of the sample estimates of the third order cross moments $\gamma_{xy}(k,l)$ and corresponding standard errors (in brackets). The length of the series is 200 and the experiment was repeated 50 times. In this case also simulation results agreed well with theoretical results. The values corresponding to cell (k--1, l-1 or k-1, l--1) were found to be dominating (-1.38) over other values (the maximum is only 0.03). Theoretically the expected value in the cells (k--1, l-1 or k-1, l--1) is 1.42 while other values should be zero. The estimates of the third order cross moments $\gamma_{yx}(k,l)$ were again found to be very close to zero for different values of k and l.

Some features of the simulation studies can be summarized as follows:

1) The third order cross moments $\gamma_{xy}(k,l)$ were found to be non-zero in some of the cells. For example, in the case of subdiagonal bivariate bilinear models, the third order cross moments will not be zero in some of the cells when k,l>0. In the case of diagonal bivariate bilinear models the

third order cross moments will not be zero when (k-k'>0, 1-0) or (1-1'>0 and k-0) for some k'-1'. Similarly in the case of superdiagonal bivariate bilinear models the third order cross moments will not be zero for some values when k<0, 1>0 or k>0, 1<0. It may be mentioned that these results are true in the case of simple bivariate bilinear models discussed above, with only one bilinear term on the right-hand-side and hence only two non-zero cells. These results could be generalized easily for more general bivariate bilinear models with several terms of similar type on the right-hand-side.

2) It is easy to prove theoretically that for simple bivariate bilinear models discussed above all the third order cross moments $\gamma_{yx}(k,1)$ will be zero for different values of k and 1. The same conjecture has been drawn through simulation studies and it was found that for all three bivariate bilinear models discussed above the third order cross moments $\gamma_{yx}(k,1)$ are very close to zero for different values of k and 1.

3) The standard errors of the estimates in all three cases were of the same magnitude. We have also carried out some individual simulations and it was discovered that for series of moderate length (say between 100-250 observations) and values of the $\beta \geq 0.5$ the pattern was distinguishable but it may not be distinguishable when the length of the series is small or the value of β is quite low. The other simulation results are more or less similar to those for univariate bilinear models mentioned in Kumar (1986b, Ch.3).

6.3 CONCLUSIONS

It should be noted that for linear bivariate models, the third order cross moments $\gamma_{xy}(k,1)$-0 for all k and 1. Hence looking at a table of the third order cross moments one can easily distinguish bivariate bilinear models from linear bivariate models (assuming the distribution of noise is Gaussian and series x_t had been prewhitened). Again, one can also distinguish between Tables 2, 3, and 4, i.e., between diagonal, subdiagonal, and superdiagonal bivariate bilinear models by looking at these tables. It may also be possible to distinguish the lags of the particular model by carefully looking at the table.

It may be more interesting to look at the third order cross moment table for a few more general bivariate bilinear models.

Table 3. Third Order Cross Moments $\gamma_{xy}(k,l)$ for the Subdiagonal Bivariate Bilinear Model $y_t = \beta y_{t-1} x_{t-2} + a_t$ where $\beta = 0.70$, $n = 200$ and Numbers of Iteration (I) = 50.

l \ k	-3	-2	-1	0	1	2	3
-3	-0.01 (0.17)	-0.02 (0.16)	-0.00 (0.17)	0.00 (0.12)	-0.01 (0.11)	-0.01 (0.17)	0.00 (0.15)
-2	-0.02 (0.16)	0.00 (0.19)	-0.04 (0.19)	-0.00 (0.15)	-0.00 (0.13)	-0.00 (0.16)	-0.02 (0.12)
-1	-0.00 (0.17)	-0.04 (0.19)	0.01 (0.18)	0.03 (0.16)	-0.03 (0.15)	0.03 (0.19)	-0.02 (0.13)
0	0.00 (0.12)	-0.00 (0.15)	0.03 (0.16)	-0.03 (0.18)	0.04 (0.16)	-0.01 (0.15)	-0.01 (0.19)
1	-0.01 (0.11)	-0.00 (0.13)	-0.03 (0.15)	0.04 (0.16)	-0.03 (0.15)	1.35 (0.14)	-0.02 (0.13)
2	0.01 (0.17)	-0.00 (0.16)	0.03 (0.19)	-0.01 (0.15)	1.35 (0.14)	-0.00 (0.14)	0.03 (0.18)
3	0.00 (0.15)	-0.02 (0.12)	-0.02 (0.13)	-0.01 (0.19)	-0.02 (0.13)	0.03 (0.18)	0.04 (0.18)

The only non-zero theoretical values in the cell $\gamma_{xy}(1,2) = \gamma_{xy}(2,1) = 1.42$.

Table 4. Third Order Cross Moments $\gamma_{xy}(k,l)$ for the Superdiagonal Bivariate
Bilinear Model $y_t = \beta y_{t-2} x_{t-1} + a_t$ where $\beta = 0.70$, n = 200 and Numbers of
Iteration (I) = 50.

l \\ k	-3	-2	-1	0	1	2	3
-3	0.03 (0.16)	0.01 (0.11)	0.01 (0.20)	-0.01 (0.11)	0.02 (0.21)	-0.00 (0.11)	0.00 (0.17)
-2	0.01 (0.11)	0.03 (0.16)	-0.00 (0.09)	0.00 (0.20)	0.01 (0.13)	0.00 (0.15)	0.00 (0.15)
-1	0.01 (0.20)	-0.00 (0.09)	0.03 (0.14)	-0.00 (0.10)	1.37 (0.17)	0.00 (0.12)	0.03 (0.19)
0	-0.01 (0.11)	0.00 (0.20)	-0.00 (0.10)	0.01 (0.16)	0.00 (0.17)	-0.00 (0.18)	-0.00 (0.12)
1	0.02 (0.21)	0.01 (0.13)	1.37 (0.17)	0.00 (0.17)	0.04 (0.15)	0.00 (0.14)	-0.00 (0.16)
2	-0.00 (0.11)	0.00 (0.15)	0.00 (0.12)	-0.00 (0.18)	0.00 (0.14)	0.01 (0.14)	-0.00 (0.14)
3	0.00 (0.17)	0.00 (0.15)	0.03 (0.19)	-0.00 (0.12)	-0.00 (0.16)	-0.00 (0.14)	0.04 (0.15)

The only non-zero theoretical values are for (k=1, l=-1) and (k=-1, l=1) and equal
to 1.42.

ACKNOWLEDGEMENT

The author is grateful to his supervisor, Dr. I.T. Jolliffe, for his helpful comments. This work was carried out when the author was at the University of Kent at Canterbury, U.K., and is a part of his Ph.D. thesis.

REFERENCES

Granger, C.W.J. and Andersen, A.P. (1978) An Introduction to Bilinear Time Series Models, Vandenhoek and Ruprecht, Gottingen.

Kumar, K. (1986a) On the Identification of Some Bilinear Time Series Models, *Journal Time Ser. Anal.*, 7, 117-122.

Kumar, K. (1986b) Some Topics in the Identification of Time Series Models, Ph.D. Thesis, University of Kent, U.K.

Subba Rao, T. (1981) On the Theory of Bilinear Time Series Models, *Journal Roy. Stat. Soc.*, B, 43, 244-255.

Subba Rao, T. and Gabr, M.M. (1984) An Introduction to Bispectral Analysis and Bilinear Time Series Models," Lecture Notes on Statistics, Springer-Verlag.

Subba Rao, T. (1986) Statistical Analysis of Bivariate Bilinear Time Series Models, Tech. Report No. 177, UMIST.

NON-LINEAR TIME SERIES MODELLING IN POPULATION BIOLOGY: A PRELIMINARY CASE STUDY

H Tong
Mathematical Institute
University of Kent
Canterbury
Kent, CT2 7NF,
England.

Summary

By reference to a preliminary analysis of the Australian blowfly data, the paper addresses the questions of 'when', 'how' and 'what' in non-linear time series modelling. Intercourse between population dynamics and statistical analysis is emphasised throughout.

Keywords: Population dynamics, ecology, Australian blowfly, food limitation, development delay, threshold, SETAR, piecewise linearity, test for linearity, non-Gaussian data, limit cycles, model robustness.

Introduction

With the field of non-linear time series modelling currently fast expanding (vide the recent review by Tong, 1986), it seems pertinent to reflect on the following questions, which are, in our view, fundamental to the development of the subject:

(1) When is a non-linear time series model called for?

(2) How to do it?

(3)   What do we get out of it?

Through the discussion of our preliminary experiences in modelling the well- known Australian blowfly data of A.J. Nicholson, we hope to throw some light on the above issues.

Some background to population dynamics and A.J.Nicholson's data

In the 1950's, the Australian entomologist, A.J. Nicholson, conducted a series of experiments with blowflies, *Lucilia cuprina*. His laboratory data have since become classic and stimulated wide interest in population ecology. Of particular note are the cycles apparent in his population data.

The data we are going to analyse are abstracted from Brillinger et al (1980) and correspond to the bi-daily record (fig.1) of one of A.J.Nicholson's experiments extending over two years, in which a caged population of approximately 1000 blowflies was initiated with a reasonably balanced sex ratio. The caged blowflies were fed a <u>limited</u> amount (about 500 mg.) of 'ground' liver daily as the only source of protein which is necessary for egg production. Experimental evidence suggests that egg production usually ceases when daily protein intake for the female fly drops below 0.14 mg. and levels out at 10 eggs per fly per day when protein supply is plentiful. On the unrealistic

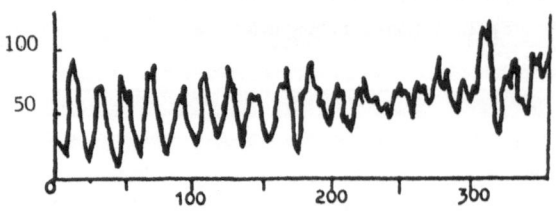

Figure 1 :   Time plot of square root of blowfly data

assumptions of absolute egalitarianism and sex equaltiy among the fly population, five hundred mg. of ground liver will maintain 3571 flies

above the minimum protein requirement for egg production. It is then
transparent that in any reasonable 'balance equation' describing the
time evolution of the population size, the number of births, as a
function of the population size, N, must peak at a certain 'critical'
value/<u>threshold</u>, say N_c, of N. Specifically, let b(N) denote the
the birth-rate as a function of N. It is then generally accepted that
an idealised birth-curve, i.e. b(N)N vs. N, would be a one- humped
curve. (Cf. Fig.2.) This is the case in, for example, the balance

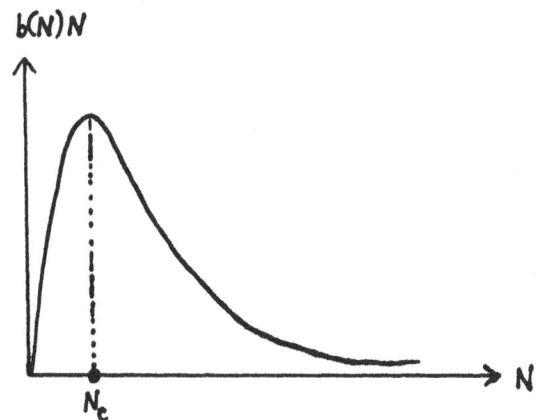

Figure 2 :   Birth Curve

equation (e.g. Oster and Ipaktchi, 1978):

$$\frac{dN(t)}{dt} = \ell(\alpha) \, b(N(t-\alpha))N(t-\alpha) - \mu \, N(t), \qquad (2.1)$$

where   N(t) = number of adults at time t,

  α = age at which an individual becomes an adult,

  $\ell(\alpha)$ = fraction of new-borns surviving to adulthood,

  b(.)   = age-averaged birth rate,

and μ = age-averaged death rate, assumed constant.

This is, in fact, a delay-differential equation, with the delay
corresponding to the so-called (biological) <u>development time</u>.

    As far as we are aware, no published results are available to
date on the fitting of the above <u>continuous-time</u> model to observed

78

data; we discuss the problem and a possible solution elsewhere. However, on a more empirical level, Readshaw and Cuff (1980) have 'fitted' a discrete time model to the blowfly data, which is in a sense analogous to model (2.1) and may be written as follows - (see also Gurney et al, 1981):

$$\Delta N(t) = f(N(t-15)) - 0.2\, N(t-1), \qquad (2.2)$$
where $\Delta N(t) = N(t) - N(t-1)$, t is an integer (presumably ≥ 15),

and
$$f(x) = \begin{cases} 10x & \text{if } x \leq 171 \\ 1795 - 0.503x & \text{if } 171 < x \leq 3569, \quad (2.3) \\ 0 & \text{if } x > 3569. \end{cases}$$

The fitting of the parameters appears to be ad hoc. and no accuracy is given. The one-humped birth-curve is approximated by the 'triangle' given by equation (2.3), (c.f. Fig. 3 with Fig. 2).

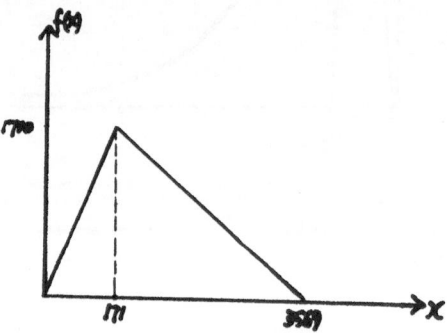

Figure 3 : Readshaw-Cuff Curve

Model (2.2) is essentially a piecewise linear model with the threshold set at 171.

Is a non-linear time series model called for?

The above biological/ecological discussion strongly suggests that a conventional linear time series model, e.g. an ARMA model, will be wide off the mark because it completely ignores the two crucial biological features:

(i)   threshold due to food limitation

and

(ii)   delay due to development time.

The next question is whether a statistical test may also alert us to the inadequacy of linearity. In view of the aforementioned discussion, one natural approach is to test the null hypothesis H_o: the blowfly data are generated by a linear time series model, against an alternative hypothesis in the direction of piece-wise linearity.

Formally, let $\{X_t : t = 0, \pm1, \pm2, \ldots\}$ denote a strictly stationary discrete-parameter time series. We may construct a test of

$$H_o : \quad b_i^{(1)} = b_i^{(2)}, \quad \text{for} \quad i = 0, 1, \ldots, p,$$

within the class of self-exciting threshold autoregressive (SETAR) models:

$$X_t = \begin{cases} b_o^{(1)} + \sum_{i=1}^{p} b_i^{(1)} X_{t-i} + e_t & \text{if} \quad X_{t-d} \leq r, \\[2mm] b_o^{(2)} + \sum_{i=1}^{p} b_i^{(2)} x_{t-i} + e_t & \text{if} \quad X_{t-d} > r, \end{cases} \tag{3.1}$$

where p and d ($\leq p$) are known positive integers and $\{e_t\}$ is a sequence of normally distributed i.i.d. random variables with zero mean and variance σ^2 and e_t is independent of X_s, $s < t$.

It may be noted that the problem is non-standard in that the nuisance parameter r is absent under H_o, which implies that the null distribution of the, say conventional, likelihood ratio statistic is unlikely to assume any simple analytic form. (See K.S. Chan and Tong, 1987). Nevertheless, we can always resort to the computing-intensive Monte Carlo method of obtaining an approximate null distribution for the likelihood ratio statistic given by

$$\lambda = \{ \hat{\sigma}_{NL}^2 / \hat{\sigma}_L^2 \}^{N/2} \tag{3.2}$$

where N denotes the sample size, $\hat{\sigma}_{NL}^2$ is the usual mean residual sum of squares under (3.1) and $\hat{\sigma}_L^2$ is that under H_o. In practice, p and d are rarely known. Monte Carlo experimentation suggests that the choice of the maximum order of the AR models is not critical for the test and a search for p is incorporated in the Monte Carlo programme over $p = 0, 1, 2,$ and over $d \leq p$. More details about the test are given in W.S. Chan & Tong (1986).

Now, turning to the blowfly data at hand, it is obvious by sight that the first half of the data set looks quite different from the second half. A more systematic methodology may be adopted using the time-dependent AR model approach described in Tong (1983, pp.255-6). Details of the analysis will be given elsewhere. For our preliminary analysis, it suffices that we do a 'split-plot' with the change point over time being set at roughly 200, and apply the test separately. Notice that the effect of the apparent trend may be reduced to some extent by the split-plot, especially for the earlier section. To complement our test, we also perform three other tests for non-linearity on the same two subsets of data. These tests are the Subba Rao-Gabr-Hinich's test, Keenan's test and Petruccelli-Davies' CUSUM test. The first one is a frequency-domain test and the latter two are time-domain tests. For a review of these tests see W.S.Chan & Tong (1986). Results of the four tests on the two subsets of the blowfly data are summarized in Table 1.

Table 1: Results of Tests Applied to Blowfly Data
(nominal 5% significance level used)

Data	Instantaneous Data Transformation	Likelihood Ratio Test (3.2)	CUSUM Test	Keenan's Test	Subba Rao-Gabr-Hinich's Test
First Part	RAW	NL (0.00)	NL (0.04)	L (0.19)	L (0.32)
	Square root	NL (0.03)	NL (0.02)	NL (0.05)	L (0.75)
	$\log_{10}$	NL (0.00)	NL (0.00)	NL (0.02)	NL (0.02)
Last Part	RAW	L (0.64)	L (0.45)	L (0.66)	L (0.12)
	Square root	L (0.87)	L (0.40)	L (0.69)	L (0.71)
	$\log_{10}$	L (0.56)	L (0.29)	L (0.74)	L (0.64)

Note: L = linear,   NL = non-linear .
       (P-values in parentheses)

On bearing in mind the earlier reported comparatively low power of Subba Rao-Gabr-Hinich's test against alternatives in the direction of piecewise linearity (see, e.g. W.S. Chan & Tong, 1986), the four tests together suggest that there is strong evidence of non-linearity in the earlier part of the blowfly data, especially in the direction of piecewise linearity (- recalling that both test (3.2) and the CUSUM test are specifically designed for this direction. It would seem interesting to investigate the robustness of all four tests w.r.t. slight trend of the data.) This leads us to conclude that, *for the earlier section of the data, an empirical non-linear time series model, which incorporates the biological features of threshold and delay is clearly called for.* As for the apparent loss of non-linearity in the later section of the data, we may suggest that the prolonged period of conditioning to a specific environment of adult-food limitation has induced a physiological change in the flies. (However, the previous remark of robustness may be more relevant here.) In fact, Brillinger et al. (1980) have reported that the population is younger at the later stages and that Nicholson's further experiment discovered that these flies could lay eggs with much less protein. Is this evidence of a selection for autogenous flies or the onset of chaos? These two suggestions might be intimately related and will be the subject of further research.

How to do it?

The problem is quite simply resolved given the present technology of time series model fitting. The fitting of a SETAR model is one of the simplest to accomplish among non-linear time series models because it is a piece-wise linear autoregressive model. Details of the fitting are omitted for the sake of brevity but may be obtained in Tong (1983), which also provides the necessary computer listing. Now, on running the said computer program, the following SETAR model may be fitted to the blowfly data on a logarithmic scale. We note that the logarithmic scale is used because

(i)   such an instantaneous transformation is commonly used in analysing population data

and

(ii) all the four tests lead to the same conclusion on this scale. However, this scale could lead to interpretational problems in some cases.

(1)   Earlier section

$$
X_t \begin{cases} 2.65 + 0.27 \, X_{t-1} + e_t \quad \text{if} \ X_{t-8} \leq 3.05 \\ (0.22) \quad (0.06) \\ \\ 0.48 + 1.40 \, X_{t-1} - 0.19 \, X_{t-2} - 0.36 \, X_{t-3} + e_t, \ \text{if o.w} \\ (0.09) \ (0.11) \qquad (0.19) \qquad (0.12) \end{cases} \tag{4.1a}
$$

where $e_t \sim N(0, 0.0148)$.

(2)   Later section:

$$
X_t = 1.17 + 0.86 \, X_{t-1} - 0.19 \, X_{t-2} + \eta_t, \tag{4.1b}
$$
$$
(0.27) \ (0.11) \qquad (0.11)
$$

where $\eta_t \sim N(0, 0.0168)$.

Here, approximate standard errors of the parameter estimates are given in parentheses. It may be noted that as remarked in the footnote of Tong (1983, p.187), the improved verion of the computer programme listed therein gives slightly more parsimonious models than those reported earlier elsewhere.

What do we get out of it?

It will be seen that besides gaining a deeper understanding of the generating mechanism of the blowfly data, a number of questions may be raised which appear to us to be interesting as well as pertinent to population dynamics.

(i)   During the earlier section, the underlying dynamics of the population (on suppressing the e_t in (4.1a)) gives rise to an asymmetric population cycle (see Fig. 4) as an intrinsic phenomenon

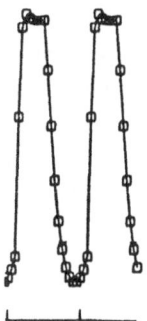

Figure 4 :   Limit Cycle

consequent upon the introduction of the threshold (at 3.05) and the
delay (equal to 8 units of time, i.e. 16 days). The cycle is in fact a
limit cycle. It may be worthwhile recalling that a stable limit cycle
enjoys the status of being robust and thus experimentally observable,
(see, e.g., Tong, 1983, Ch.2).   No extrinsic models, such as those
obtained by appending sinusoidal terms to an AR model, could hope to
share this status. It would be an understatement to say that there is
an intellectual gulf between the two paradigms.

(ii)   The statistically identified time delay of 8 units of time (i.e.
16 days) in model (4.1a) fits in quite well with the observed time to
emergence (from egg to immature adult) of between 10 and 16 days. It
may also be recorded that slight variations of the delay parameter do
not alter the other parameter estimates or the limit cycle to any great
extent — slight variations w.r.t. other parameters experience similar
effects.   In other words, there seems to exist some inbuilt
robustness/stability of the fitted model

(iii) The threshold parameter estimate of 3.05 may be translated into
a statistically critical population size of 1122 flies.   It might be
ecologically meaningful to investigate the implications of this value
in terms of efficiency of food utilization, random crowding effect,
etc.   Interestingly, large perturbations of the threshold parameter
leads to the replacement of the limit cycle by limit points.

(iv)   The mildly double-humped plateau of the cycles shown in Fig.4 might deserve further investigation. Some theoretical ecologists seem to be quite excited about such double-humps (see, e.g. Readshaw & van Gerwen, 1983). Will the introduction of a secondary delay (say corresponding to the development time from egg to pupa of approximately 6 days) in the SETAR model accentuate the humps into the fork-like feature evident in the early section of the observation record? If so, does that make ecological sense? These questions may merit further investigations.

$$U = X_t \qquad V = X_{t-8}$$

for fitted SETAR

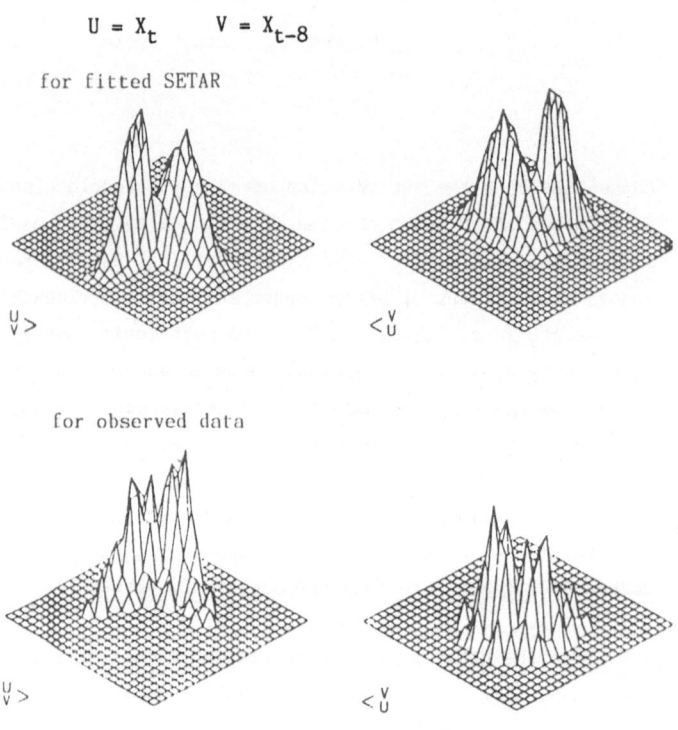

for observed data

Figure 5 :   Bivariate Distribution of (X_t, X_{t-8})

(v)   As far as the statistical goodness of fit of the model (4.1a) is concerned, the gain over linear Gaussian models is typified by Fig.5, in which the distinct non-normality of the observed distribution of (X_t, X_{t-8}), is captured by the fitted distribution through the model.

It may be noted that the non-linear 1-step-ahead post-sample (of size 60) prediction based on (4.1a) reduces the root-mean-square error of a linear 1-step-ahead post-sample prediction based on the AR model selected by minimum AIC by approximately 15%.

(vi)   For reasons mentioned earlier, the linear Gaussian model (4.1b) fitted to the later section of the data can only be considered tentative. For example, further investigation into the possibility of an onset of chaos might be of substantial interest in view of the extreme importance of the new concept.

6. Discussion

The data appears to exhibit 'heteroscedasticity' over its dynamic range in that the conditional variance may not be a constant. To some extent, this may be built into the SETAR models (e.g. 4.1a) by allowing the term e_t to have difference variances depending on the regimes (i.e. X below or above the threshold).

Having put forward a SETAR model as a reasonble empirical non-linear time series model with some theoretical underpinning for the blowfly data, the question may still be raised as to the validity of a 'discontinuous' model. K.S. Chan and Tong (1986) addressed this question by first embedding the discontinuous SETAR as the limit of a net of smooth threshold autoregressive (STAR) models. In place of model (4.1a) the following STAR model may then be entertained:

$$X_t = (2.51 + 0.34\, X_{t-1} - 0.03\, X_{t-2}) + (-2.32 + 1.10\, X_{t-1}$$
$$\quad\;\; (0.31)\;\;\; (0.12)\qquad\;\; (0.07)\qquad\quad (0.35)\;\;\; (0.17)$$

$$-0.49\, X_{t-2})\;\; \Phi\; ((X_{t-8} - 3.103)\,/\, 0.084) + e_t, \qquad (5.1)$$
$$(0.15)\qquad\qquad\qquad (0.025)\;\;\; (0.045)$$

where Φ is the c.d.f. of $N(0,1)$ and $e_t \sim N(0,0.021)$. In view of the
$$(0.002)$$

small scale parameter (at 0.084), it seems reasonable to suggest that it is not illegitimate to idealise $\Phi((x-3.103)/0.084)$ to a Heaviside function with the 'jump' at 3.103. Evidence suggests that this idealisation improves the (statistical) goodness of fit.

One of the limitations of the preliminary models reported here is that they cannot explain the trend as an <u>intrinsic</u> property. It is also clear that no models can be complete without taking into account the egg counts, death counts, etc.

Acknowledgement

I am grateful to the participants at the NSF–NBER Seminar in Time Series Analysis at the Southern Methodist University, Dallas, Texas, U.S.A., in October 1986, for comments on the paper.

References

Brillinger, D.R., Guckenheimer, J., Guttorp, P.E. & Oster, G. (1980) "Empirical modelling of population time series data: the case of age and density dependent vital rates". <u>Lectures on Mathematics in the Life Sciences</u>, (Amer.Math.Soc.), Vol.13, 65–90.

Chan, K.S. & Tong, H. (1986) "On estimating thresholds in autoregressive models" J.Time Series Anal., Vol.7, 179–190.

Chan, K.S. & Tong, H. (1987) "A survey of the statistical analysis of univariate threshold autoregressive models" – to appear in <u>Advances in Statistical Analysis and Statistical Computing: Theory and Applications</u>, Ed. R.S. Mariano, JAI Press Inc.,U.S.A.

Chan, W.S. & Tong, H. (1986) "On tests for non-linearity in time series analysis" To appear in <u>Journal of Forecasting</u>.

Gurney, W.S.C., Blythe, S.P. & Nisbet, R.M. (1981) "Reply to a letter from J.L.Readshaw" <u>Nature</u>, Vol.292, 178.

Oster, G. & Ipaktchi, A. (1978) "Population Cycles" <u>Theor.Chem: Periodicities in Chem. & Biol.</u> (ed. H.Eyring & D. Henderson), New York: Academic Press, 111–132.

Readshaw, J.L. & Cuff, W.R. (1980) "A model of Nicholson's blowfly

cycles and its relevance to predation theory" J.Anim.Ecol.,
Vol.49, 1005-1010.

Readshaw, J.L. & van Gerwen, A.C.M. (1983) "Age-specific survival,
fecundity and fertility of the adult blowfly,   ,
in relation to crowding, protein food and population cycles"
J.Anim.Ecol., Vol.52.

Tong, H. (1983) "Threshold models in non-linear time series analysis"
Lectures Notes in Statistics, No.21. New York: Springer-Verlag.

Tong, H. (1986) "Non-linear time series models of regularly sampled
data: a review" An invited paper presented at the First World
Congress of the Bernoulli Society of Mathematical Statistics and
Probability, held in September, 1986, in Tashkent, USSR.
Proceedings published by VNU Science Press, The Netherlands.

THE AKAIKE INFORMATION CRITERION IN THRESHOLD MODELLING:

Some Empirical Evidences

W.K. LI
DEPARTMENT OF STATISTICS
UNIVERSITY OF HONG KONG

Abstract The performance of the Akaike information criterion in threshold modelling is studied using simulation. Particular attention is paid to the effects of autoregressive parameters, the maximum order entertained, and the choice of possible candidates for the delay and threshold parameters in the procedure.

1. Introduction

Non-linear time series has recently received a lot of attention. That this is so, perhaps, is not too surprising as in real-life situations many time series are clearly non-linear. Linear models, such as the auto-regressive moving-average models, can provide first order approximations to such series. There are, however, certain features in non-linear series that cannot be reproduced effectively by linear approximations. For example, it would be difficult for Gaussian autoregressive moving-average models to model phenomena such as limit cycles and time irreversibility. One is referred to Tong (1980), Tong and Lim (1980) and the book by Tong (Tong, 1983) for more discussions on this matter.

In the above works, a new type of time series models called the threshold models is proposed. One distinguished feature of this new type of models is their flexibility. Phenomenon such as limit cycles which is an important feature in cyclical data, can be modelled effectively.

As an example of threshold model, consider the univariate series $\{x_t\}$ generated by

$$x_t = a_0^{(1)} + a_1^{(1)} x_{t-1} + \ldots + a_{p_1}^{(1)} x_{t-p_1} + e_t^{(1)} \text{ if } x_{t-d} > r, \quad (1.1)$$

and

$$x_t = a_0^{(2)} + a_1^{(2)} x_{t-1} + \ldots + a_{p_2}^{(2)} x_{t-p_2} + e_t^{(2)} \quad \text{otherwise,}$$

where $\{a_j^{(i)}\}$, $j = 0,1,\ldots,p_i$, $i = 1,2$, are constant parameters and $\{e_t^{(i)}\}$, $i = 1,2$, are independent identically distributed random variates. Here r is the threshold parameter and d is the delay parameter. More formally, $\{x_t\}$ is called a self exciting threshold autoregressive (SETAR) model.

More details can be found in Tong (1980) and Tong and Lim (1980). It
suffice, for the purpose of this paper, to use (1.1) as a working
definition for threshold autoregressive models.

Of course, in actual practice all the parameters of (1.1) would be
unknown and would have to be estimated. Some model identification and
diagnostic tools have been proposed in the above mentioned works. A very
important tool is the Akaike information criterion (AIC). Suppose now that
there are M competing models, then AIC(m), the AIC for the mth model is
given by

$$AIC(m) = -2\ln \text{ (maximum likelihood)} +$$

$$2 \text{ (number of independent parameters).} \qquad (1.2)$$

The model which minimizes AIC(m) over $m = 1,2,\ldots,M$, would be picked as the
right model.

The AIC has been used extensively in autoregressive moving-average
modelling (see for example, Akaike, 1969). In that case, its property is
pretty well known. For example, Shibata (1976) has found that the AIC is
not consistent in picking the correct autoregressive order.

In the threshold case, not only the two autoregressive orders in (1.1)
but also the delay and threshold parameters are of interest. Just how good
the minimum AIC estimates these parameters is not well understood. It is
therefore important to give the minimum AIC procedure a critical
evaluation, at least in some simple cases. Theoretical results, however,
seem difficult to obtain and thus simulations are used. Despite
limitations to such an approach, it is believed that enough light could
still be cast on the general situation. Teräsvirta and Luukkonen (1985)
study the related problem of choosing between linear and threshold
autoregressive models using AIC and BIC.

2. Method of Estimation

For completeness of presentation the estimation procedure is summed up
as follow. Tong (1983) suggested essentially the same procedure.

Let L be the maximum order for each of the two autoregressive processes
in (1.1). Let $\Gamma = \{1,2,\ldots,T\}$ and $R = \{r_1,r_2,\ldots,r_S\}$ be the sets of
potential candidates of d and r respectively. Suppose the length of
realization of $\{x_t\}$ is N. The procedure is as follows:
(1) Fix d and r at d_0 and r_0. Let the tentative orders be $1 \leqslant k_1, k_2 \leqslant L$.
    Partition $\{x_1,x_2,\ldots,x_N\}$ into two disjoint vectors using the rule

$$x_j \in X_1 \quad \text{if } x_{j-d} \geq r,$$
$$x_j \in X_2 \quad \text{if } x_{j-d} < r. \tag{2.1}$$

The first N_{d_0} X_j's would be discarded, where $N_{d_0} = \max(d_0, L)$. Form correspondingly, the simultaneous linear equations

$$X_1 = A_1\theta_1 + e_1,$$
$$X_2 = A_2\theta_2 + e_2, \tag{2.2}$$

where

$$X_i = (x_{j_1}^{(i)}, x_{j_2}^{(i)} \ldots, x_{j_{n_i}}^{(i)})^T,$$

$$e_i = (e_{j_i}^{(i)}, \ldots, e_{j_{n_i}}^{(i)})^T,$$

$$\theta_i = (a_o^{(i)}, \ldots, a_{k_i}^{(i)})^T,$$

and

$$\begin{bmatrix} 1 & x_{j_1-1}^{(i)}, & \ldots, & x_{j_1-k_i}^{(i)} \\ 1 & x_{j_2-1}^{(i)}, & \ldots, & x_{j_2-k_1}^{(i)} \\ \cdot & \cdot & \ldots & \cdot \\ \cdot & \cdot & \ldots & \cdot \\ \cdot & \cdot & \ldots & \cdot \\ 1 & x_{j_{n_i}-1}^{(i)}, & \ldots, & x_{j_{n_i}-k_i}^{(i)} \end{bmatrix}$$

for $i = 1, 2$.

For Gaussian $\{e_t^{(1)}\}$ and $\{e_t^{(2)}\}$, the least squares procedure yields approximate maximum likelihood estimates and the Householder transformation (Kennedy and Gentle, 1980) can be used to diagonalize A_i, $i = 1, 2$.

Denote by $\hat{k}_i$ the minimum AIC estimates of p_i, $i = 1, 2$. Then $\hat{k}_i$ is such that

$$AIC(\hat{k}_i) = \min_{0 < k_i < L} \{n_i \ell n(||\hat{e}_i(k_i)||^2/n_i) + 2(k_i+1)\},$$

where $||\cdot||$ denotes the Euclidean norm. Let AIC $(d_0, r_0) = AIC(\hat{k}_1) + AIC(\hat{k}_2)$.

(2) Fix d at d_0 and vary r over R. Choose $\hat{r}$ such that

$$AIC(d_0,\hat{r}) = \min_{r \in R} \{AIC(d_0,r)\}.$$

(3) Now vary d over Γ. For each d, the $AIC(d,\hat{r})$ is normalized by dividing it by $N - N_d$. The minimum AIC estimate $\hat{d}$ of d is that d which satisfies

$$AIC(\hat{d}) = \min_{d \in \Gamma} \{AIC(d,\hat{r})/(N-N_d)\}.$$

The parameters $\hat{r}$, $\hat{k}_1$ and $\hat{k}_2$ obtained corresponding to $\hat{d}$ are the final AIC estimates of r, p_1, and p_2.

3. **The Empirical Study**

In this paper, the threshold autoregressive model

$$x_t = a \, x_{t-1} + e_t, \quad \text{if } x_{t-1} > 0,$$
$$x_t = b \, x_{t-1} + e_t, \quad \text{if } x_{t-1} < 0, \tag{3.1}$$

will be considered. The interest is on the minimum AIC estimates of p_1, p_2, r and d. In particular, the effects of the parameters (a,b) and the sets Γ and R are considered. $\{e_t\}$ are assumed to be $N(0,1)$ distributed. The International Mathematical and Statistical Library (IMSL) is used to generate the normal random variates and to perform the Householder transformation. It is assumed that the constant parameters are known to be zero. The matrices A_i, i = 1,2 and the AIC's are adjusted accordingly. In all cases, there are 100 independent realizations each of length 200. It is further assumed that p_1, $p_2 \neq 0$.

(a) **General Study.**

In order to obtain a general picture of the performance of the AIC, an extensive number of models (3.1) is considered. Table 1 sums up results under six different cases. Here L = 3, R = $\{-1,0,1\}$ and $\Gamma = \{1,2,3\}$. There are 81 possible models to be selected. It is obvious that many of these can be grouped together. Moreover, if the delay parameter is not correctly estimated than there is no point to study the threshold parameter. Thus, the outcomes of the estimation prcedure are grouped into six cases as follows:

(I)   Choosing the correct delay;

(II) Case (I) plus choosing the correct threshold;

(III) Case (II) plus choosing at least one autoregressive order correct;

(IV) Case (II) plus choosing correctly both autoregressive orders. That
       is, the correct model;

(V)   Both autoregressive orders correct, regardless of the delay and the
       threshold parameters;

(VI) Case (V) plus the correct delay parameter.

These six cases seems to be of major interests and comprehensive.

There are several distinct features which made Table 1 very instructive. Firstly, the estimation of the delay parameter appears to be very
effective. For the first ten models the relative frequencies of correctly
estimating d are about 0.9 or higher. The frequencies of correctly
selecting both the d and r parameters are also high. These, however, may
be due to the choice of Γ and R and will be studied further. Secondly,
looking down the columns for cases (II) and (III), it can be seen that
given the $\hat{d}$ and $\hat{r}$ are correct the relative frequency is almost 1 that at
least one of the autoregressive orders is also correct, It seems, however,
more difficult to have everything right, as is indicated in case (IV).
Thirdly, if the two parameters (a,b) are close to each other, the AIC
performs poorly. The further apart the parameters (a,b) the better is the
estimation result. The first four and the last four models illustrate this
point. Fourthly, if the threshold parameter is neglected then it can be
seen under the last two columns that estimation results improve greatly.
For the first four models, the improvement is less dramatic than the last
four models. This further suggests that if (a,b) are close then estimation
of the threshold parameter is difficult. Considering that the corresponding probability for picking the right model for the linear first order
autoregressive case is about 0.72 (Shibata, 1976), the performance of the
AIC in selecting correctly the full model is perhaps not too satisfactory.

(b)   The Effects of L and Γ.

Table 2 gives the estimation results when L = 6 for four of the models
in Table 1. R and Γ remain the same as in (a). For reason of comparison,
the same sequences of realization as those in Table 1 are used. The
performance of the minimum AIC drops in all cases except the first. The
estimation of the delay parameter remains very "robust" against the
increase in L. Table 3 sums up results when Γ is increased to
{1,2,3,4,5,6} and L = 6. The last five cases are only mildly affected.
The last two models feel most of this negative effect. The frequencies of
picking the right d also drop although a relative frequency of 0.75 or
above can be quite acceptable.

(c)   The Effect of R.

Tables 4 and 5 sum up results for the four models in (b) when R = {-1,-.5,0,.5,1} and {-1,-.7,0,.7,1} respectively.   L and Γ remain the same as in study (a).   Again for the sake of comparison, the same realizations as in (a) are used.

Comparing Tables 4 and 5 with Table 1, it can be seen that the performance of the AIC decreases sharply for all cases involving the threshold estimate, namely cases (II), (III) and (IV).   Even in the first two models, the frequency of selecting the right model is only about 30 percent in Table 4. $\hat{d}$ remains as robust as before, so are casese (V) and (VI) where the threshold estimate is not considered.   It can also be seen that results in Table 5 are better than those in Table 4.   In Table 5, the other possible candidates are further away from the true threshold.   This suggests that the choice of possible thresholds has a significant effect on $\hat{r}$ and the final outcome of the AIC.

This poses some sort of a dilemma to the practical researcher, since often it is necessary to have a large number of candidates r_1, r_2, ..., r_s in order to include (or approximate) the true threshold.   But if the number of possible thresholds is too many it appears that one would have difficulties finding the right one.   Chan and Tong (1986) study the problem of estimating the threshold by introducing smoothness into the model (1.1).   However, their study assumes known order and delay.   Further research in this direction is important and necessary.   In addition, note that in both Tables 4 and 5, the chance of selecting at least one of p_1 and p_2 correct given $\hat{d}$ and $\hat{r}$ are already correct, is still close to 1.

4.   Conclusions

In this paper, simulation results suggest that the estimation of the delay parameter by the minimum AIC in threshold modelling can be very effective.   In addition, given the delay and threshold are chosen right, then very likely one of the autoregressive orders will also be right.   In practice, model diagnostic techniques can perhaps take care of the remaining order.   Values of the autoregressive parameters on both sides of the threshold can also greatly affect the AIC results.   It would be of interest to compare the performance of the Bayesian information criterion (Schwarz, 1978) with that of the AIC.   This problem is currently under investigation by the author.   Clearly, there are still a lot of important theoretical and empirical questions to be answered in the identification of non-linear time series models.

Table 1

Frequencies of cases (I) to (VI). $\Gamma = \{1,2,3\}$, $L = 3$, $R = \{-1,0,1\}$.

Model (a,b)	(I)	(II)	(III)	(IV)	(V)	(VI)
1. (-.7, .7)	100	97	95	69	70	70
2. (.7,-.7)	100	95	92	55	56	56
3. (-.5, .5)	100	86	83	60	66	66
4. (.5,-.5)	100	95	92	53	57	57
5. (-.3, .3)	95	62	59	35	53	53
6. (.3,-.3)	97	67	61	38	53	53
7. (.2, .8)	85	56	56	36	50	49
8. (-.2,-.8)	99	77	73	49	60	60
9. (.8, .2)	95	72	58	38	59	58
10. (-.8,-.2)	100	72	69	43	55	55
11. (.3, .5)	66	22	20	11	41	26
12. (-.3,-.5)	53	25	22	16	60	34
13. (.5, .3)	62	21	19	11	44	29
14. (-.5,-.3)	66	26	26	13	47	30

Table 2

Frequencies when $L = 6$, $\Gamma = \{1,2,3\}$, and $R = \{-1,0,1\}$.

(a,b)	(I)	(II)	(III)	(IV)	(V)	(VI)
(-.5, .5)	100	84	76	48	51	51
(.5,-.5)	100	91	88	46	49	49
(-.2,-.8)	97	66	63	34	43	43
(.8, .2)	88	50	46	27	41	39

Table 3

Frequencies when L = 6, Γ = {1,2,3,4,5,6}, and R = {-1,0,1}.

(a,b)	(I)	(II)	(III)	(IV)	(V)	(VI)
(-.5, .5)	99	83	75	47	50	50
(.5,-.5)	100	91	88	46	49	49
(-.2,-.8)	75	64	60	32	40	40
(.8, .2)	77	47	40	23	35	34

Table 4

Frequencies when R = {-1,-.5,0,.5,1}.

Model	(I)	(II)	(III)	(IV)	(V)	(VI)
(-.5, .5)	100	48	45	29	59	59
(.5,-.5)	100	53	51	31	53	53
(-.2,-.8)	97	39	38	25	48	48
(.8, .2)	92	33	31	19	58	55

Table 5

Frequencies when R = {-1,-.7,0,.7,1}.

Model	(I)	(II)	(III)	(IV)	(V)	(VI)
(-.5, .5)	100	71	66	48	65	65
(.5,-.5)	100	76	73	42	58	58
(-.2,-.8)	98	52	50	34	54	54
(.8, .2)	98	36	32	21	58	57

References

Akaike, H. (1969). Fitting autoregressive models for prediction. Ann.
  Inst. Statist. Math., 21, 243–247.

Chan, K.S. and H. Tong (1986). On estimating thresholds in autoregressive
  models. J. Time Series Analysis, 7, 179–190.

Kennedy, W.J. and Gentle, J. (1980). Statistical Computing. New York:
  Marcel Dekker.

Schwarz, G. (1978). Estimating the dimension of a model. Annals of
  Statistics, 6, 461–464.

Shibata, R. (1976). Selection of the order of an autoregressive model by
  Akaike's information criterion. Biometrika, 63, 117–126.

Tervasvirta, T. and R. Luukkonen (1985). Choosing between linear and
  threshold autoregressive models. Time Series Analysis: Theory and
  Practice, 7 (ed. O.D. Anderson), Amsterdam: North Holland.

Tong, H. (1980). A view on non-linear time series model building. Time
  Series (ed. O.D. Anderson), Amsterdam: North-Holland.

Tong, H. and Lim, K.S. (1980). Threshold autoregression, limit cycles and
  cyclical data (with discussion). J. Roy. Statist. Soc., B, 42, 245–292.

Tong, H. (1983). Threshold models in non-linear time series analysis.
  Springer Lecture Notes in Statistics, 21.

NONLINEAR TIME SERIES ANALYSIS
FOR DYNAMICAL SYSTEMS OF CATASTROPHE TYPE

Loren Cobb
Department of Biometry
Medical University of South Carolina
Charleston, SC 29425

Shelemyahu Zacks
Department of Mathematical Sciences
State University of New York
Binghamton, NY 13901

1 Introduction

This chapter is concerned with a class of nonlinear time series models inspired by catastrophe theory, a branch of differential topology. The nonlinear models of catastrophe theory offer interesting modes of behavior that are not found in the usual linear models of time series analysis. For example, the cusp catastrophe model exhibits a phenomenon called "bistability," which means that the state variable has two attracting equilibrium points, separated by a repelling equilibrium point. Figure 1.1 shows a bistable time series generated by a stochastic cusp catastrophe model. The two attracting equilibria are at ±1 (indicated by dotted lines), while the repelling equilibrium is at zero.

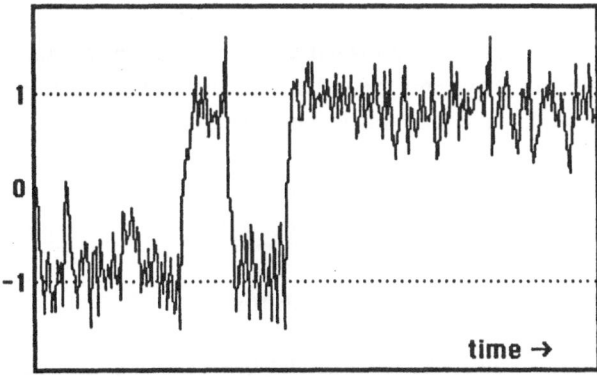

Figure 1.1 A time series generated by a bistable stochastic catastrophe model.

Note the tendency of the process to remain for a while within the domain of attraction of one attracting equilibrium, before finally moving past zero and into the domain of attraction of the other.

Catastrophe models share with most nonlinear models the characteristic that their qualitative behavior (e.g. the number of attracting equilibria) depends on their parameters. Thus slowly changing one parameter of a catastrophe model can result in the sudden appearance or disappearance of pairs of equilibria. Characterizing these qualitative modes of behavior is, in fact, the subject matter of catastrophe theory.

The models of catastrophe theory are not statistical models, but we present a method for deriving statistical time series models from the canonical catastrophes. The resulting models would most naturally take the form

$$X_{t+1} = \theta_0 + \theta_1 X_t + \theta_2 X_t^2 + \theta_3 X_t^3 + \dots + \theta_d X_t^d + U_t, \tag{1.1}$$

were it not the fact that such models are, in general, neither stationary nor ergodic. This is a major problem for the statistical analysis of such models, since most methods assume that their models are both stationary and ergodic. As Tjøstheim (1986) has remarked, the task of finding nonlinear models satisfying these assumptions is far from trivial! In Section 2 of this chapter we show that the source of the difficulty for models of the form (1.1) lies in the existence of a periodic orbit in the deterministic version of these models, and possibly uncontrolled oscillations. In Section 3 we present a modification of the model that is both stationary and ergodic, while still retaining the essential characteristics of the canonical catastrophe model from which it was ultimately derived. We consider in some detail a special case of the modified cubic time series model, and analyze the maximum likelihood estimators of its parameters. This article is designed to show that the array of research problems that is opened up when nonlinear time series analysis is seen from the perspective of dynamical systems theory is vast, and that straightforward applications of results which are available in the statistical literature is not always possible.

2 Polynomial Time Series Models

2.1 Nonlinear Dynamical Systems

In this section we describe the varieties of behavior exhibited by the class of first-order time series models that are nonlinear in their state variables, namely models of the general form

$$X_{t+1} = f(\theta, X_t) + U_t, \tag{2.1.1}$$

where f is a function that is nonlinear in its second argument, and $\{U_t\}$ is a sequence of independent and identically distributed normal random variables. Note that if f is linear in its first argument, the parameter vector θ, then in statistical terminology it is a "linear model." In order to avoid confusion in the usage of the terms "linear" and "nonlinear", we shall call the models discussed here "time series models for nonlinear dynamical systems." For example,

$$X_{t+1} = a + bX_t + cX_t^2 + U_t \qquad (2.1.2)$$

is a linear model for a nonlinear dynamical system, whereas

$$X_{t+1} = e^{-at}X_t + U_t \qquad (2.1.3)$$

is a nonlinear model for a linear dynamical system. Our interest lies with models of the former kind, not the latter. Further, we shall restrict our attention in this section to the behavior of the deterministic parts of these models. The stochastic behavior will be reserved for Section 3.

The subclass of models of the form (2.1.1) in which f is a polynomial in the state variable is of central importance in the description of the varieties of behavior that can be exhibited by the entire class. This importance originates in the theorems of catastrophe theory, which use the polynomial models as the canonical members of equivalence classes of nonlinear models. However, the subject matter of catastrophe theory is the classification of deterministic systems in continuous time, not stochastic systems in discrete time such as (2.1.1).

The major topic of this section is the classification of nonlinear dynamical systems as seen in catastrophe theory, and the additional considerations implied by the use of discrete time in time series models.

2.2 First-Order Stochastic Dynamical Systems

A *first-order* dynamical system in discrete-time is characterized by the fact that the rate of change of the state variable is dependent only on the current value, and not upon previous values:

$$\Delta x_t = f(x_t)\Delta t, \qquad (2.2.1)$$

where Δ is the forward difference operator (with interval Δt) defined by

$$\Delta x_t = x_{t+\Delta t} - x_t. \qquad (2.2.2)$$

When this dependence is modified by the presence of additive stochastic noise, as in

$$\Delta X_t = f(X_t)\Delta t + U_t, \quad U_t \sim N(0, \sigma^2 \Delta t), \tag{2.2.3}$$

then we have a model which is an autoregressive nonlinear dynamical system. The connection between (2.2.3) and the AR(1) model (as it is known in the time series analysis literature) is evident if we let $\Delta t = 1$ and $f(x) = \mu + \theta x$, which yields:

$$X_{t+1} = \mu + (1+\theta)X_t + U_t. \tag{2.2.4}$$

Both equations (2.2.3) and (2.2.4) are examples of *stochastic difference equations*.

A first-order (non-stochastic) dynamical system in continuous time is characterized by the differential equation:

$$\dot{x}_t = f(x_t), \tag{2.2.5}$$

where $\dot{x}$ is the time derivative of x. The corresponding *stochastic differential equation* is:

$$dX_t = f(X_t)dt + \sigma dW_t, \tag{2.2.6}$$

where W_t is the standard Wiener Process (Liptser & Shiryayev, 1977, pp. 82-88). The close relationship between this system and the discrete-time system (2.2.3) becomes apparent when we identify U_t with $\sigma\Delta W_t$: the continuous-time version is just a discrete-time system with an infinitesimal Δt (see Stroyan & Bayod, 1986, for a rigorous derivation of stochastic differential equations from infinitesimal difference equations).

The theory of statistical estimation and inference for the continuous-time nonlinear systems (2.2.6) is quite well developed (Liptser & Shiryayev, 1977, 1978) in comparison to the discrete-time nonlinear systems (2.2.3). The goal of this chapter is to show that statistical estimation and inference for discrete-time nonlinear systems is both feasible and practical, provided care is taken in the initial specification of the statistical model. We restrict this exposition, however, to the special case of cubic models.

2.3 Attractors and Repellors in First-Order Systems

The class of nonlinear systems presents a dramatically greater variety of behavior than the class of linear systems. Consider first the equilibrium structure. The equilibrium points of (2.2.2) and (2.2.5) are defined as the points x such that $\Delta x = 0$ or $\dot{x} = 0$, respectively, *i.e.*

$\{\ x:\ f(x) = 0\ \}.$ (2.3.1)

Clearly, if the system is linear and nontrivial then there is at most one equilibrium point. A nonlinear system, by contrast, has as many equilibria as there are roots of $f(x) = 0$.

An equilibrium point is said to be *repelling* if the state variable eventually departs from any arbitrarily small neighborhood of the point, never to return. Conversely, an equilibrium point is said to be *attracting* if the state variable approaches the point asymptotically as $t \to \infty$. For example, the nonlinear dynamical system defined by

$\dot{x} = \theta(a\text{-}x)(b\text{-}x)(c\text{-}x), \quad \theta < 0.$ (2.3.2)

has three equilibria, two attracting and one repelling, as depicted in Figure 2.1:

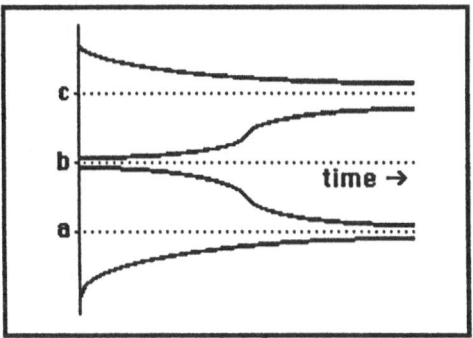

Figure 2.1 Trajectories of the system defined by (2.3.2). The equilibria at a and c are attracting, but the equilibrium at b is repelling.

The *domain of attraction* for an attractor is the set of points in the state space that are attracted to it. Thus the interval $(-\infty, b)$ is the domain of attraction for a in the above figure, while (b, ∞) is the domain of attraction for c.

Now consider the discrete-time dynamical system

$\Delta x = -\theta x^3 \Delta t, \quad \theta > 0.$ (2.3.3)

This system clearly has an attracting equilibrium at $x = 0$, but what is its domain of attraction? From Figure 2.2 we see that there is an interval $(-A, A)$ within which the system moves asymptotically towards the origin. However, given an initial position outside this interval the system begins oscillating about the origin with an ever-increasing amplitude. We shall call such destructive oscillations an *explosion*.

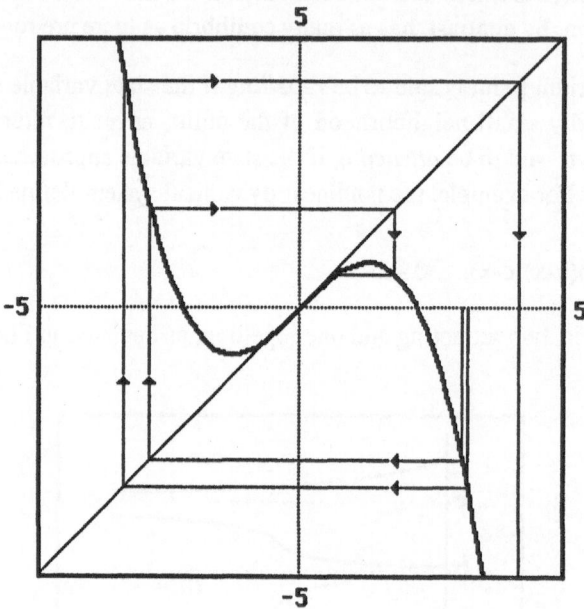

Figure 2.2 A "cobweb" graph for two trajectories of the system defined by Equation (2.3.3), from two initial positions: 3.1 and 3.2, with θ = 0.2 and Δt = 1. The set {±√10} is a period-2 repellor, while the origin is an attractor.

The boundary of the domain of attraction for the origin in the system (2.3.3) depends upon θ. It consists of the set of points A such that if x_t = A then $x_{t+\Delta t}$ = -A. Equation (2.3.3) is easily solved for these points:

$$A = \pm(2/\theta)^{1/2} \Delta t. \qquad (2.3.4)$$

Notice that the set {-A, A} constitutes a *periodic orbit* (of period 2) for this system. It is easy to see that this orbit is repelling, since any deviation from the orbit will cause the system either (a) to move towards the attractor at the origin, or (b) to begin an explosion. A periodic orbit that is attracting is also known as a *limit cycle*.

Repelling (or attracting) periodic orbits are not found in linear dynamical systems. The periodic behavior of the linear second-order system

$$x_{t+1} = x_t - \theta x_{t-1}, \qquad\qquad (2.3.5)$$

is neither attracting nor repelling, since any deviation from a periodic orbit of this system results only in the establishment of a new periodic orbit of the same frequency but different amplitude and phase. Thus the periodic orbits of nonlinear systems constitute a form of behavior that is qualitatively different from anything seen in linear dynamical systems.

These two examples illustrate the two most important characteristics of nonlinear dynamical systems which are missing in linear dynamical systems:

     (1) *Multiple attracting and repelling equilibrium points.*
     (2) *Attracting and repelling periodic orbits.*

To describe the entire range of behavior of first-order nonlinear dynamical systems, it will be necessary to introduce some additional terminology. In these notes we follow in part the survey by Zeeman (1982). Suppose now that the state space S is $\Re^n$. As before, we are concerned with either differential or difference systems:

$$dx = f(x)dt \quad or \quad \Delta x = f(x)\Delta t. \qquad\qquad (2.3.6)$$

The collection of solutions $\{x_t\}$ for $dx = f(x)dt$ constitutes a *flow*, which is a function $\psi : S \to S$ such that $\psi_s(x_t) = x_{s+t}$, or, stated a little more elegantly,

$$\psi_s(\psi_t) = \psi_{s+t}. \qquad\qquad (2.3.7)$$

For systems in discrete time ψ is often called a *map* instead of a flow, and is defined only on multiples of Δt. Note that $\psi_{\Delta t}(x_t) = x_t + f(x_t)\Delta t$. The fixed points of $\psi_{\Delta t}$ are the equilibria of the system.

An *orbit* passing through a point x is the curve described by $\psi_t(x)$, for all t. In the discrete-time case, the orbit is the set of discrete points encountered by $\psi_t(x)$. An orbit is periodic if $\psi_\tau(x_t) = x_t$ for some $0 < \tau < \infty$ and all points x in the orbit at any t. A *period-n orbit* of a discrete-time system consists of points x such that $\psi_{n\Delta t}(x) = x$, but $\psi_{m\Delta t}(x) \neq x$ for all $0 < m < n$.

A point x is *non-wandering* if for every neighborhood N of x and all $t \in \Re$ (all $t/\Delta t \in Z$ if discrete-time), there is an $s > t$ such that $N \supseteq \psi_s(N)$. The set of all such points is the *non-wandering set* Ω. The non-wandering set of a nonlinear system may include equilibria as well as periodic orbits and non-periodic orbits.

A subset Λ of the state space is said to be *attracting* if there is a closed neighborhood N of Λ such that $\psi_t(N) \subset N$ for all $t > 0$, and $\bigcap_{t>0} \psi_t(N) = \Lambda$. Four kinds of attractors are known, in general:

    i.   Point attractors (equilibria).
    ii. Periodic attractors (limit cycles).
    iii. Toroidal attractors (quasi-periodic orbits).
    iv. Strange attractors (chaotic orbits).

Strange attractors are particularly important. A flow or map which contains a strange attractor typically exhibits an extreme sensitivity to initial conditions, such that orbits that begin close together rapidly diverge. This gives an unpredictable character to the motion, which explains why it is called "chaos." *All of the nonlinear discrete-time catastrophe models that we consider here are capable of exhibiting chaos.*

The cubic dynamical system

$$\dot{x} = \theta_0 + \theta_1 x + \theta_2 x^2 + \theta_3 x^3 \qquad (2.3.8)$$

may have from one to three equilibria; the number depends upon the parameter vector θ. In the case of three equilibria, as in Figure 2.1, there are typically two possibilities for the non-wandering set: an attractor surrounded by two repellors, or a repellor surrounded by two attractors.

The discrete-time case,

$$\Delta x = (\theta_0 + \theta_1 x + \theta_2 x^2 + \theta_3 x^3)\Delta t, \qquad (2.3.9)$$

has a much more complicated non-wandering set. In addition to three equilibria, located at the roots of the cubic polynomial, the non-wandering set can contain periodic, toroidal, and chaotic orbits. An example of a chaotic orbit is shown in Figure 2.3, on the next page. Note its characteristic pseudo-random trajectory.

Recall that the cubic dynamical systems (2.3.8) and (2.3.9) have one or three equilibria depending upon the parameters of the model. In general the structure of the non-wandering set of a nonlinear dynamical is very dependent upon the parameters. As a parameter is smoothly varied, the non-wandering set may undergo sudden changes in composition, *e.g.* from one to two attracting equilibria, or from a toroidal attractor to a strange attractor, *etc.* A *catastrophe* (in the broadest sense) is a transition in which an attractor disappears from the non-wandering set. The cubic dynamical systems considered here are among the simplest nonlinear models which can exhibit catastrophes.

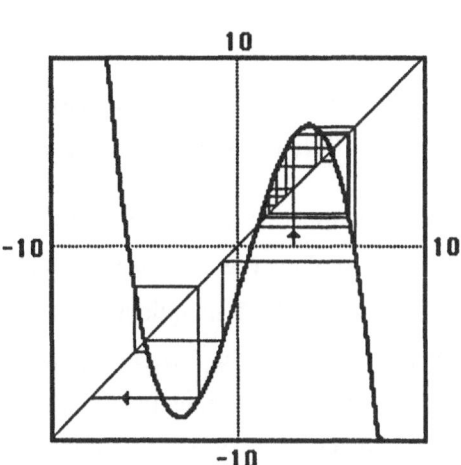

Figure 2.3 A cobweb graph that exhibits chaotic motion. The system is governed by $\Delta x = -0.09(x-5)(x-1)(x+5)\Delta t$, with an initial position at $x = 3$, and $\Delta t = 1$.

A sufficient condition for chaos is the existence of a period-3 orbit (Li & Yorke, 1975). The boundary of the chaotic domain for "bimodal" maps (a category which includes the polynomial systems considered here) has been analyzed exhaustively by MacKay & Tresser (1985).

A cubic dynamical system of the form (2.3.8) with $\theta_3 < 0$ may or may not have any period-2 orbits. If it does, then one period-2 orbit is a repellor which encloses all of the equilibria of the system. Lemma 2.1 (below) gives sufficient conditions for the existence of such a period-two orbit. We provide the proof in complete detail because of the importance of the existence of this orbit for the censored model presented in Section 3.

Lemma 2.1: Let $x_{t+1} = \psi_1(x_t)$, where

$$\psi_1(x) = \alpha_0 + \alpha_1 x + \alpha_2 x^2 + \alpha_3 x^3, \quad \alpha_3 < 0. \tag{2.3.10}$$

If this system has at least one attracting equilibrium point, then it also has at least one period-2 orbit, and the largest such orbit encloses all fixed points of ψ_1.

Proof of Lemma 2.1: Let $\psi_2(x) = \psi_1(\psi_1(x))$. Thus ψ_2 is a ninth degree polynomial, say $\psi_2(x) = \beta_0 + \beta_1 x + ... + \beta_9 x^9$. Notice that $\beta_9 = \alpha_3^4 > 0$, and that all the fixed points of ψ_1 are also fixed points of ψ_2. Obviously, $\psi_1(x) - x = 0$ has at most three real roots. We distinguish between two cases:

Case 1: $\psi_1(x) - x = 0$ has one real root of multiplicity 1;

Case 2: $\psi_1(x) - x = 0$ has three real roots (not necessarily distinct).

Case 1: If r is the unique attracting fixed point of ψ_1, then $|\psi_1'(r)| < 1$. Indeed, as is easy to check, if $\psi_1'(r) < -1$ then r is not an attractor. Moreover, $\psi_1'(r) < 0$; otherwise, since $\psi_1(x) \to \infty$ as $x \to \infty$, the assumptions of Case 1 are violated. By simple differentiation, we find that $\psi_2'(x) = \psi_1'(x)\psi_1'(\psi_1(x))$. Hence $\psi_2'(r) = [\psi_1'(r)]^2$, and $0 < \psi_2'(r) < 1$. Furthermore, $\psi_2'(x)$ is a continuous function, such that $\psi_2'(x) = 9\alpha_3^4 x^8 + O(x^7)$ as $x \to \infty$. Hence, $\exists s > r$ such that $\psi_2'(x) > 1$ for all $x > s$. Let t be the point such that $\psi_2(t) = t$. Thus t is a fixed point of ψ_2 which is not a fixed point of ψ_1. Consequently $\{t, \psi_2(t)\}$ is a period-2 orbit of the system defined by (2.3.10).

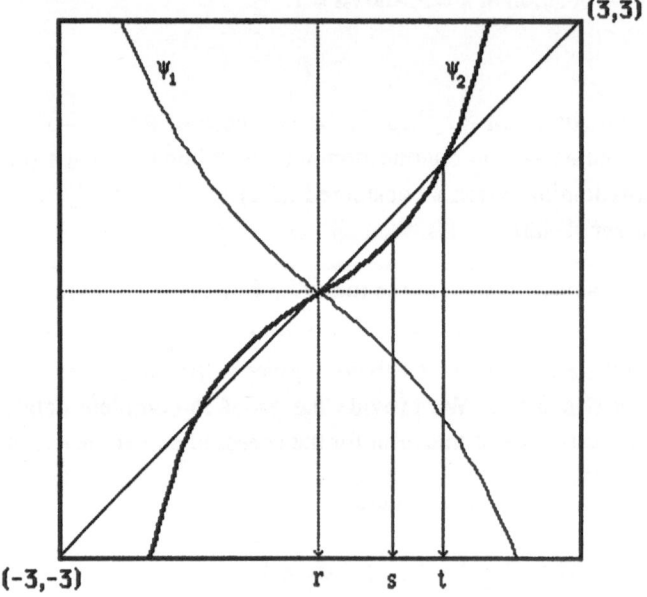

Figure 2.4 The functions ψ_1 and ψ_2 are graphed here in thin and thick lines, respectively, with $\psi_1(x) = -0.8x + 0.1x^3$. The origin is the fixed point of ψ_1. Note that $\psi_1'(0) = -0.8$, and that ψ_2 has three fixed points. This illustrates Case 1 of Lemma 2.1.

Case 2: Let the fixed points of ψ_1 be r_1, r_2 and r_3, with $r_1 \le r_2 \le r_3$ (see Figure 2.5). There must exist a point, say s, such that $s < r_1$ and $\psi_1(s) = r_3$, because $\alpha_3 < 0$. The line labelled A in Figure 2.5 extends from (s,r_3) to (r_3,r_3). But $\psi_2(s) = r_3$ also, since $\psi_2(s) = \psi_1(\psi_1(s)) = \psi_1(r_3) = r_3$. Now the fact that $\beta_9 > 0$ implies that there must exist a point, say t, such that $t < s$ and $\psi_2(t) = t$. The point $(t,\psi_2(t))$ is labelled C in the figure. A similar argument using the line labelled B in the figure shows the existence of the point labelled D. The points C and D constitute a period-2 orbit, within which lie all the fixed points of ψ_1.

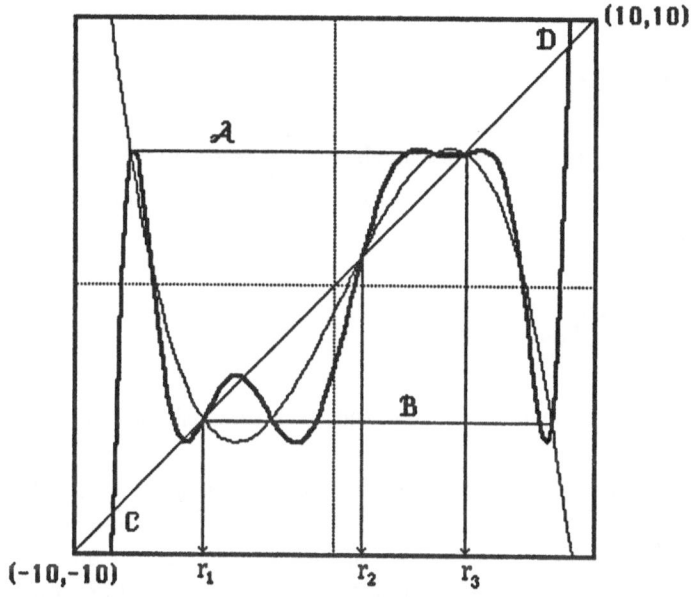

Figure 2.5 The functions ψ_1 and ψ_2 are graphed in thin and thick lines, respectively, with $\psi_1(x) = x - 0.04(x-5)(x-1)(x+5)$. The period-2 orbit is {C,D}. This illustrates Case 2.

(End of Proof)

3 Nonlinear Time Series Analysis

3.1 Introduction

Three main questions are connected with the topic of nonlinear time series analysis. One is related to the choice of model, the second with the estimation of the parameters of the chosen model, and the third with assessing the statistical properties of the estimators. The question of which nonlinear model to choose depends on the type of data under consideration, and on the characteristics of the model. In the present paper, we focus attention on cubic models of order one, with or without covariates. This class of models is derived from the catastrophe models presented in Section 1. The general form of the cubic time series (CTS) model without covariates is:

$$\Delta X_t = \theta_0 + \theta_1 X_t + \theta_2 X_t^2 + \theta_3 X_t^3 + \sigma U_t, \quad (t = 0, 1, ...), \tag{3.1.1}$$

where $\{U_t: t=0,1,...\}$ is a stationary stochastic process with mean zero for all t and a specified correlation function. The general form of the CTS with k covariates is similar to (3.1.1), with

$$\theta_i = \alpha_i + \sum_{j=1}^{k} \beta_{ij} Y_{tj}, \quad (i = 0, 1, 2, 3), \tag{3.1.2}$$

where $\{Y_t.\}$ is a given sequence of (non-random) k-dimensional covariates.

Unfortunately, models (3.1.1) and (3.1.2) are ill-behaved for the purposes of time series analysis, since they are not, in general, stationary. The nonstationarity is due to the finiteness of the domain of attraction—outside the domain of attraction the system exhibits destructively increasing oscillations. There are several possible ways to overcome this problem: (1) the cubic polynomial can be approximated with a piece-wise linear function, or (2) the model can be modified so that it is polynomial on a finite interval surrounding its equilibria, but changes to a different form outside this interval. We have chosen the latter alternative, so as to preserve the close connection between these models and the polynomial models of catastrophe theory. The former alternative is also viable, as in the piecewise linear SETAR models of Tong (1980, 1983), but these models lack the close connection to catastrophe theory that we are seeking.

In this section we present a class of "censored" polynomial models for time series analysis. These models are ergodic, and manage to preserve almost all of the desirable features of discrete time catastrophe models. We devote the remainder of this section to a thorough analysis of a special one-parameter model with no covariates which, it is hoped, faithfully reflects the essential difficulties that will be found in the more general models.

3.2 Properties of a Special Cubic Time Series Model

Consider the CTS given by

$$X_0 \equiv 0,$$

$$\Delta X_t = -\omega X_t^3 + \sigma U_t, \quad (t = 0, 1, 2, ...),$$

$(3.2.1)$

with $\omega > 0$, $\sigma > 0$, and $\{U_t\}$ an *i.i.d.* sequence of standard normal random variables. In the deterministic case ($\sigma = 0$) there is one attracting equilibrium, at the origin. By Lemma 2.1 there is also a period-2 orbit at $\{-A, A\}$, where $A = \sqrt{2/\omega}$. All points in the interval $(-A, A)$ are attracted to the origin, while all points outside of $(-A, A)$ are repelled towards $\pm\infty$. Thus the origin is an attractor, while the periodic orbit is a repellor.

Let $g(x,t)$ denote the PDF of X_t. From the Markovian properties of the CTS (3.2.1) we obtain

$$g(x,1) = \varphi(x/\sigma)/\sigma,$$

$$g(x,t) = \int_{-\infty}^{\infty} \varphi([x-y+\omega y^3]/\sigma)g(y,t-1)dy, \quad (t \geq 2),$$

$(3.2.2)$

where $\varphi(z)$ denotes the PDF of the standard normal distribution. One can prove, by induction on t, that $g(-x,t) = g(x,t)$, for all $t \geq 1$, therefore the distribution of X_t is symmetric around the origin. Hence, all the odd moments of X_t are zero for all $t \geq 0$.

The variance of X_t satisfies the difference equation:

$$\text{Var}[X_0] = 0,$$

$$\Delta\text{Var}[X_t] = \sigma^2 - 2\omega E[X_t^4] + \omega^2 E[X_t^6], \quad (t=1,2,...).$$

$(3.2.3)$

This shows that $\text{Var}[X_t]$ may grow very fast with t, if ω is sufficiently large. Indeed, when t=1 we have $X_1 \sim N(0,\sigma^2)$ and

$$\Delta\text{Var}[X_1] = \sigma^2 - 6\omega\sigma^4 + 15\omega^2\sigma^6.$$

$(3.2.4)$

This difference is greater than σ^2 whenever $\omega > 6/15\sigma^2$. Moreover, if ω is large then $A = \sqrt{2/\omega}$ is small, and with high probability the stochastic process will rapidly exit the domain of attraction $(-A, A)$, and begin destructive oscillations towards $\pm\infty$. The following lemma shows that the time required for the CTS (3.2.1) to exit the domain of attraction is stochastically smaller than a random variable having a geometric distribution.

Lemma 3.1: Let X_t satisfy (3.2.1), and let τ_A be the first exit time from the domain of attraction (A,-A), where $A = \sqrt{2/\omega}$. Thus

$$\tau_A = \text{least } t{\geq}1 \text{ such that } |X_t| \geq A. \tag{3.2.5}$$

Then

$$P\{\,\tau_A > t\,\} < [\,\Phi(\,2^{3/2}/\sigma\sqrt{\omega}\,)\,]^t, \quad (t{\geq}1), \tag{3.2.6}$$

where $\Phi(z)$ is the standard normal integral.

Proof of Lemma 3.1: Define the "defective" cumulative distribution function

$$H(x,t) = P\{\,X_t \leq x, \tau_A \geq t\,\}, \quad (t \geq 1). \tag{3.2.7}$$

Notice that $H(x,1) = \Phi(x/\sigma)$. Because the CTS has the Markov property,

$$H(x,t) = \int_{-A}^{A} \Phi(\,[x-y+\omega y^3]/\sigma\,)h(y,t-1)dy, \quad (t{\geq}2), \tag{3.2.8}$$

where $h(x,t) = \partial H(x,t)/\partial x$. In particular, for all $t \geq 2$,

$$H(A,t) = \int_{-A}^{A} \Phi(\,[A-y+\omega y^3]/\sigma\,)h(y,t-1)dy. \tag{3.2.9}$$

It is easy to verify that

$$\sup_{-A \leq y \leq A} \{\,A-y+\omega y^3\,\} = 2A. \tag{3.2.10}$$

Hence, from the monotonicity of $\Phi(z)$, we obtain

$$H(A,t) \leq \Phi(2A/\sigma) \int_{-A}^{A} h(y,t-1)dy \tag{3.2.11}$$

$$\leq \Phi(2A/\sigma)\,[H(A,t-1) - H(-A,t-1)].$$

Notice that, from definition (3.2.7),

$$P\{\tau_A > t\} = H(A,t) - H(-A,t) \tag{3.2.12}$$

$$< H(A,t)$$

$$\leq \Phi(2A/\sigma)\,P\{\tau_A > t{-}1\}$$

$$\leq [\,\Phi(2A/\sigma)\,]^{t-1}\,P\{\tau_A > 1\}$$

$$\leq [\,\Phi(2A/\sigma)\,]^{t-1}\,[\,2\Phi(A/\sigma){-}1\,]$$

$$\leq [\,\Phi(2A/\sigma)\,]^{t}$$

$$= [\,\Phi(2^{3/2}/\sigma\sqrt{\omega})\,]^{t}. \qquad \textbf{(End of Proof)}$$

Thus the first exit time τ_A is stochastically smaller than a random variable τ^* having a geometric distribution:

$$P\{\tau^* = \tau\} = [\,1{-}\Phi(2^{3/2}/\sigma\sqrt{\omega})\,]\,[\,\Phi(2^{3/2}/\sigma\sqrt{\omega})\,]^{t-1}, \quad (t{>}0). \tag{3.2.13}$$

It follows that the expected first exit time from the domain of attraction satisfies

$$E[\tau_A] \leq 1/[\,1{-}\Phi(2^{3/2}/\sigma\sqrt{\omega})\,]. \tag{3.2.14}$$

If $\sigma\sqrt{\omega}$ is large then this expectation is small and the first exit will occur rapidly. Conversely, if $\sigma\sqrt{\omega}$ is small then the expected time to the first exit is very large.

In a similar manner we can show that, once the process (3.5) is outside the domain of attraction, the time until the first entrance back into $(-A,A)$ is stochastically greater than the geometric random variable having a mean of $1/[1/2 - \Phi(-2^{3/2}/\sigma\sqrt{\omega})]$.

3.3 A Censored Cubic Time Series Model

In order to avoid the undesirable behavior of the unmodified cubic time series model, we now consider a version which is "censored" within the domain of attraction, namely

$$Y_0 \equiv 0,$$

$$Y_{t+1} = [\,Y_t - \omega Y_t^3 + \sigma U_t\,]_{-A}^{+A}, \quad (t = 0, 1, 2, \ldots), \tag{3.3.1}$$

where $A = \sqrt{2/\omega}$, $\omega > 0$, $\sigma \geq 0$, and

$$[x]_{-A}^{+A} = \begin{cases} A, & x > A \\ x, & -A \leq x \leq A. \\ -A, & x < -A \end{cases} \quad (3.3.2)$$

Like the uncensored model (3.2.1), the censored model is Markov, but unlike (3.2.1) it is ergodic, with positive recurrent states. Oscillations of the censored CTS (3.3.1) can occur at the points $\pm A$, which are the boundaries of the domain of attraction. Moreover, if $\tau^*(A)$ denotes the number of trials needed to return to the domain of attraction, given that $Y_t = \pm A$, then $\tau^*(A)$ has a geometric distribution

$$P\{ \tau^*(A) = k \mid Y_t = \pm A \} = \Psi(A)(1-\Psi(A))^{k-1}, \quad (k = 1, 2, ...), \quad (3.3.3)$$

regardless of t, where

$$\Psi(A) = \Phi(2^{3/2}/\sigma\sqrt{\omega}) - 1/2. \quad (3.3.4)$$

The transition cumulative distribution function for the censored process has two mass points, at $x = \pm A$. It is given by

$$p(x, y; \omega, \sigma) = P\{ Y_{t+1} \leq x \mid Y_t = y \} \quad (3.3.5)$$

$$= \begin{cases} 0 & \text{if } x < -A, \\ \Phi([x-y+\omega y^3]/\sigma) & \text{if } -A \leq x \leq A, \\ 1 & \text{if } x \geq A. \end{cases}$$

Furthermore, the cumulative distribution function of Y_t, $G(x,t; \omega,\sigma)$, can be determined recursively, according to these equations:

$$G(x,1; \omega,\sigma) = \begin{cases} 0 & \text{if } x < -A, \\ \Phi(x/\sigma) & \text{if } -A \leq x < A \\ 1 & \text{if } A \leq x, \end{cases} \quad (3.3.6)$$

$$G(x,t; \omega,\sigma) = I\{x \geq A\} + I\{-A \leq x < A\} \, G(-A,t-1;\omega,\sigma) \, \Phi([x-A]/\sigma)$$

$$+ \int_{-A}^{A} g(y,t-1;\omega,\sigma)\Phi([x-y+\omega y^3]/\sigma)dy + (1-G(A,t-1;\omega,\sigma))\Phi([x+A]/\sigma)],$$

$$(3.3.7)$$

where I{·} is an indicator function and $g(y,t; \omega,\sigma) = \partial G(y,t; \omega,\sigma)/\partial y$.

The stationary distribution of Y_t, *i.e.* $G^*(x) = \lim_{t \to \infty} G(x,t)$, can be approximated by the stationary distribution of a discrete Markov Chain obtained by partitioning (-A,A) into N subintervals, and including two additional states corresponding to the points -A and A.

3.4 Maximum Likelihood Estimators for the Censored Model

In this section, we present an algorithm for the determination of the maximum likelihood estimators of the parameters (ω,σ) of the censored cubic time series model (3.3.1).

Let $y^{[T]} = [y_1, ..., y_T]$ be a vector of T observed consecutive sample values of the stochastic process Y_t, where Y_t is governed by (3.3.1). Define the random variables

$$m_T = \min\{y_1, ..., y_T\},$$

$$M_T = \max\{y_1, ..., y_T\},$$

$$\omega^*_T = \min\{2/m_T^2, 2/M_T^2\}.$$

Since $|Y_t| \le (2/\omega)^{1/2}$ for all t with probability one,

$$\omega \le \omega^*_T. \tag{3.4.1}$$

Thus, the likelihood function of (ω,σ), given the observed vector $y^{[T]}$, satisfies $L(\omega,\sigma; y^{[T]}) = 0$, for all $\omega > \omega^*_T$. For $\omega \in (0,\omega^*_T)$, the likelihood function can be written as

$$L(\omega,\sigma; y^{[T]}) = \sigma^{-T} \exp[-\sum_{t=0}^{T-1} (y_{t+1} - y_t + \omega y_t^3)^2/2\sigma^2], \tag{3.4.2}$$

with $Y_0 \equiv 0$. The value of the likelihood function at ω^* can be determined as follows. Define

$$K_1 = \{t: y_t = -(2/\omega^*_T)^{1/2} \text{ and } t \in \{0,1, ... ,T\}\},$$

$$K_2 = \{t: y_t = +(2/\omega^*_T)^{1/2} \text{ and } t \in \{0,1,...,T\}\},$$

$K_3 = (\neg K_1) \cap (\neg K_2)$.

Let $J_i = |K_i|$ be the cardinality of K_i. Notice that $J_i \geq 0$ for $i = 1$ and 2, but $J_1 + J_2 \geq 1$. Both J_1 and J_2 are positive if $m_T = -M_T$. Thus the likelihood at ω^*_T is

$$L(\omega^*, \sigma; y^{[T]}) = \prod_{t \in K_1} \Phi(\, U(\omega^*_T, t-1)/\sigma\,) \prod_{t \in K_2} \Phi(\, V(\omega^*_T, t-1)/\sigma\,) \; \sigma^{-J_3} \exp[\, -Q(\omega^*_T)\,], \qquad (3.4.3)$$

where

$$U(\omega, t) = -(2/\omega)^{1/2} - y_t + \omega y_t^3,$$

$$V(\omega, t) = +(2/\omega)^{1/2} + y_t - \omega y_t^3, \quad \text{and}$$

$$Q(\omega) = \sum_{t \in K_3} (y_{t+1} - y_t + \omega y_t^3)^2.$$

Notice that if K_i is empty, then the corresponding product in (3.4.3) is equal to 1 by definition.

The log-likelihood function $\lambda(\omega, \sigma; y^{[T]})$ is, for $\omega < \omega^*_T$,

$$\lambda(\omega, \sigma; y^{[T]}) = -T \log \sigma - \sum_{t=0}^{T-1} (y_{t+1} - y_t + \omega y_t^3)^2 / 2\sigma^2. \qquad (3.4.4)$$

At $\omega = \omega^*_T$ we have

$$\lambda(\omega^*_T, \sigma; y^{[T]}) = \sum_{t \in K_1} \log \Phi(\, U(\omega^*_T, t-1)/\sigma\,) + \sum_{t \in K_2} \log \Phi(\, V(\omega^*_T, t-1)/\sigma\,)$$

$$- J_3 \log \sigma - Q(\omega^*_T)/2\sigma^2. \qquad (3.4.6)$$

Note that $U(\omega^*_T, t-1) \leq 0$ for all $t \in K_1$, and $V(\omega^*_T, t-1) \leq 0$ for all $t \in K_2$, since $|y_t| \leq (2/\omega)^{1/2}$ for all t.

The *Maximum Likelihood Estimator* (MLE) of (ω, σ) is the point in $[0, \omega^*_T] \times (0, \infty)$ at which $\lambda(\omega, \sigma; y^{[T]})$ attains its supremum. In order to determine the MLE, we find first a point (ω, σ) for which $\lambda(\omega, \sigma; y^{[T]})$ attains its supremum over the subset $(0, \omega^*_T) \times (0, \infty)$, *i.e.* we first ignore the possibility that $\omega = \omega^*_T$. It is easy to check that the *unique* point maximizing (3.4.4) is

$$\hat{\omega}_T = \min\{ \omega^*_T, -\sum_{t=0}^{T-1} y_t^3 \Delta y_t \Big/ \sum_{t=0}^{T-1} y_t^6 \} \quad \text{if } \sum_t y_t^3 \Delta y_t < 0,$$

$$= 0 \qquad\qquad\qquad\qquad \text{if } \sum_t y_t^3 \Delta y_t \geq 0,$$

$$\hat{\sigma}_T = \Big[\sum_{t=0}^{T-1} (y_{t+1} - y_t + \omega_T y_t^3)^2 / T \Big]^{1/2}. \tag{3.4.7}$$

The maximal likelihood

$$\lambda_1 = \lambda(\hat{\omega}_T, \hat{\sigma}_T; y^{[T]}) \tag{3.4.8}$$

is then compared with the boundary likelihood

$$\lambda_2 = \sup_{0<\sigma<\infty} \lambda(\omega^*_T, \sigma; y^{[T]}). \tag{3.4.9}$$

If $\lambda_1 \geq \lambda_2$ then the MLE is $(\hat{\omega}_T, \hat{\sigma}_T)$, otherwise the MLE is (ω^*_T, σ^*_T), where σ^*_T is the value of σ for which λ_2 is attained in (3.4.9).

Furthermore, due to the continuity of the distributions $G_t(y)$, for $|y| < A$,

$$P\{ (J_1 > 1) \cup (J_2 > 1) \} = P_{\omega,\sigma}\{\omega^* = \omega\}. \tag{3.4.10}$$

Hence, for all $y^{[T]}$ such that the event $\{J_1 > 1\} \cup \{J_2 > 1\}$ occurs,

$$\lambda(\omega, \sigma; y^{[T]}) = 1 \quad \text{if } \omega = \omega^*_T,$$

$$= 0 \quad \text{otherwise.}$$

We show now that, if $J_1 + J_2 < N$, there exists a unique point σ^*_T, in $(0, \infty)$, for which $\lambda(\omega^*_T, \sigma; y^{[T]})$ attains its supremum. Indeed, partially differentiating (3.4.6) with respect to σ, and equating the derivative to zero, we find that σ^*_T is the root of the equation

$$\sigma^2 = Q(\omega^*_T)/J_3 + (\sigma/J_3) \Big\{ \sum_{t \in K_1} |U(\omega^*_T, t-1)| \, \phi(U(\omega^*_T, t-1)/\sigma) / \Phi(U(\omega^*, t-1)/\sigma)$$

$$+ \sum_{t \in K_2} |V(\omega^*_T, t-1)| \, \phi(V(\omega^*_T, t-1)/\sigma) / \Phi(V(\omega^*, t-1)/\sigma) \Big\}. \tag{3.4.11}$$

We see immediately that $(\sigma^*_T)^2 > Q(\omega^*_T)/J_3$. Thus, when $J_3 = 0$, there is no finite ML estimator of σ. Moreover, if $H^*(\sigma)$ denotes the r.h.s. of (3.4.11), we see that $H^*(\sigma)$ is continuous, that $H(0) = Q(\omega^*_T)/J_3$, and that

$$H^*(\sigma) \leq Q(\omega^*_T)/J_3 + \sigma(J_1+J_2) / \{J_3(\pi\omega^*_T)^{1/2} \Phi(-(2/\sigma\omega^*_T)^{1/2}) \}. \qquad (3.4.12)$$

As $\sigma \to \infty$ the increase of the r.h.s. of (3.4.12) is approximately linear, while the l.h.s. of (3.4.11) is quadratic. Hence there exists a unique root, σ^*_T, of (3.4.11).

3.5 Properties of the MLE for the Censored Model

We have seen in Section 3.3 that, with probability one, almost all realizations of the censored model (3.3.1) attain the boundaries $\pm A$ infinitely often. Thus sooner or later, with probability one, $J_1+J_2 > 1$, and $\omega^*_T = \omega'$, where ω' is the true value of ω. Thus the **MLE of ω is strongly consistent,** *i.e.*

$$\lim_{T\to\infty} P_{\omega',\sigma}\{ MLE_T(\omega) = \omega' \} = 1, \qquad (3.5.1)$$

where $MLE_T(\omega)$ denotes the MLE of ω, given $y^{[T]}$. The likelihood function of σ, under ω^*_T, *i.e.* $L(\omega^*_T,\sigma; y^{[T]})$, satisfies the regularity conditions for consistency and asymptotic normality of $MLE_T(\sigma)$, as can be checked from (Basawa & Rao, 1980, pp. 122-125). The asymptotic distribution of $MLE_T(\omega)$ is, however, not normal.

3.6 Discussion

It is noteworthy that the our maximum likelihood estimator (3.4.7) is, in many cases, identical to the least-squares estimator

$$LSE_T(\omega) = -\sum_{t=0}^{T-1} y_t^3 \Delta y_t / \sum_{t=0}^{T-1} y_t^6, \qquad (3.6.1)$$

and also closely resembles the maximum likelihood estimator for the stochastic differential equation (recall that $\Delta t = 1$)

$$dy_t = -\omega y_t^3 dt + \sigma dw_t, \qquad (3.6.2)$$

namely

$$\text{MLE}_T(\omega) = -\int_0^T y_t^3 dy_t / \int_0^T y_t^6 dt. \tag{3.6.3}$$

We have overcome the nonstationarity problem by censoring the model at its largest period-2 orbit, and studied the properties of maximum likelihood estimators in a simple special case of a censored model. We have not discussed here the asymptotic distribution of the MLE of ω. It is clear that the classical theory does not apply here because of the non-ergodicity in the non-censored case. In the censored case simulations have shown that the asymptotic distribution of the MLE of ω is, apparently, a mixture of a normal distribution and a singular mass centered at ω.

In the present paper we have considered only cubic polynomial models, inspired by the cusp catastrophe model. More general polynomial models are available (see Cobb & Zacks, 1985). Extending the results and methods of catastrophe theory to discrete time stochastic systems is, however, quite problematic. The classification theorems of elementary catastrophe theory (Poston & Stewart, 1978) apply only to continuous time deterministic systems that have non-wandering sets which contain only discrete points. The problem is not the distinction between stochastic and deterministic differential equations—that is easily dealt with by restating the theory in terms of perturbations of the deterministic part of the stochastic differential equation. The fundamental problem is that one-dimensional discrete-time nonlinear systems do not, in general, have non-wandering sets that contain exclusively discrete points. Even the simple quadratic model $\Delta x = a+bx+cx^2$ has a chaotic domain within its parameter space. Thus one cannot use the theorems of catastrophe theory to claim the existence of an exhaustive equivalence relation on discrete-time models of low codimension. However, one can use catastrophe theory to motivate the selection of polynomial models for nonlinear time series analysis, on the basis that the family of polynomial time series models does at least exhibit generic bifurcations of its isolated equilibria—this much we can recover from catastrophe theory. Elementary catastrophe theory is silent on the question of bifurcations to periodic and chaotic orbits, which requires further mathematical research.

In summary, this is our approach: First, we claim that the family of polynomial time series models is sufficiently rich in its range of qualitative behavior to justify its use as the fundamental class of models for nonlinear time series analysis. Second, we show that all the isolated equilibria of a cubic polynomial model lie within an enclosing periodic orbit. Third, we show that by censoring the model at the enclosing periodic orbit, we obtain a stationary and ergodic process. Fourth, we derive estimators for the parameters this model that converge to the correct values with probability one.

4 References

Basawa, I.V. & Rao, B.L.S. (1980) **Statistical Inference for Stochastic Processes.** New York: Academic Press.

Cobb, Loren, and Zacks, Shelemyahu (1985) "Applications of catastrophe theory for statistical modeling in the biosciences," **Journal of the American Statistical Association, 78,** 124-130.

Li, T.Y. and Yorke, J.A. (1975) "Period three implies chaos," **American Mathematical Monthly, 82,** 985-992.

Liptser R.S. & Shiryayev, A.N. (1977) **Statistics of Random Processes I: General Theory.** New York: Springer-Verlag.

Liptser R.S. & Shiryayev, A.N. (1978) **Statistics of Random Processes II: Applications.** New York: Springer-Verlag.

MacKay, Robert S. and Tresser, C. (1985) "Boundary of chaos for bimodal maps of the interval," preprint, Institute of Mathematics, University of Warwick, Coventry, England.

Poston, Timothy, and Stewart, Ian N. (1978) **Catastrophe Theory and its Applications.** London: Pitman.

Stroyan, K.D., and Bayod, J.M. (1986) **Foundations of Infinitesimal Stochastic Analysis.** Amsterdam: North-Holland.

Thompson, J.M.T. (1982) **Instabilities and Catastrophes in Science and Engineering.** New York: Wiley.

Tjøstheim, Dag (1986) "Estimation in nonlinear time series models," **Stochastic Processes and their Applications, 21,** 251-273.

Tong, Howell, & Lim, K.S. (1980) "Threshold autoregression, limit cycles, and cyclical data (with discussion)," **Journal of the Royal Statistical Society, Series B, 42,** 245-292.

Tong, Howell (1983) **Threshold Models in Nonlinear Time Series Analysis.** Lecture Notes in Statistics #21. New York: Springer-Verlag.

Zeeman, E.C. (1982) "Bifurcation and catastrophe theory," **Contemporary Mathematics, 9,** 207-272.

NONLINEAR PROCESSING WITH Mth-ORDER SIGNALS

F.J. Bugnon and R.R. Mohler
Department of Electrical and Computer Engineering
Oregon State University
Corvallis, OR 97331 (USA)

1. INTRODUCTION

Signal processing techniques perform statistical treatments over data sequences that are random in nature, in order to recover or enhance some hidden information. The classical way to summarize a random sequence is to characterize it by its moments (mean, variance, etc.). Nevertheless, no particular random sequence can perfectly be described this way, and the only manner not to lose any information is to keep the whole original data sequence! Hence, the actual trend is to directly process raw measurements, taking advantage of processing capabilities of modern computers.

In this chapter, nonlinear functions of data sequences are generated in an analogous form to the statistical moments. From an origianl data sequence, Mth-order sequences are built through convolution techniques, and these new sequences are to be processed instead of the initial sequences. For a linear observation equation, a generating function can be defined to conveniently relate the Mth-order measurement sequences to their signal counterparts.

The direction-finding problem is a straightforward illustration of the advantages of the Mth-order signals. The example shows how eigenstructures are generated from Mth-order signals that readily apply to the direction-finding problem, and it produces independent solutions with gains either in resolution or in the number of sensors, to solve problems with coherent sources.

2. NONLINEAR SEQUENCE GENERATOR

2.1 Mth-Order Signals

A vector r of random components $r_{i,1}$ is measured at time t, or, in the discrete case, at time t_k. Let the Mth-order signals $r_{i,1}(t)$ be defined as

$$r_{i,M}(t) - r_{i,M-1}(t) * r_{i,1}(t) , \tag{1}$$

where * is the convolution operator. The Fourier transform yields the corresponding Mth-order vector R_M as

$$R_M^T - [R_{1,1}^M, \ldots, R_{j,1}^M, \ldots, R_{d,1}^M] . \tag{2}$$

The upper-case characters denote the Fourier transform of the corresponding lower-case symbols, and the arguments have been dropped for short. It follows that the Mth-order spectral-density matrix is defined as

$$D_{RM}(\omega) - E[R_M(\omega)\bar{R}_M(\omega)] , \tag{3}$$

where ω is the frequency variable and the overbar is the Hermitian transpose, or as

$$D_{RM} - E \begin{bmatrix} (R_{1,1}\bar{R}_{1,1})^M & (R_{1,1}\bar{R}_{2,1})^M & . & . & . \\ (R_{2,1}\bar{R}_{1,1})^M & (R_{2,1}\bar{R}_{2,1})^M & . & . & . \\ . & . & & . & . \\ . & . & . & & . \\ . & . & & . \end{bmatrix} . \tag{4}$$

2.2 The Δ Product

In (4), each component is raised to power M; to facilitate matrix notation and algebra, a new matrix operator needs to be defined. Let the operator Δ be defined as

$$R_M(\omega) - R_1(\omega)^{\Delta M} . \tag{5}$$

For example, when M - 2, each matrix element is simply squared.

Δ performs a component to component multiplication, i.e.,

$$A\Delta B - C <=> c_{ij} - a_{ij} \cdot b_{ij} , \tag{6}$$

where it is implicit that matrices A, B, and, of course, C are of identical dimensions.

Applied to p by q matrices, useful identities are

$$A \Delta B = B \Delta A \ ,$$

$$A \Delta (B \Delta C) = (A \Delta B) \Delta C \ ,$$

$$\overline{A \Delta B} = \overline{A} \Delta \overline{B} \ , \tag{7}$$

$$(A+B) \Delta C = A \Delta C + B \Delta C \ ,$$

$$I \Delta I = I \ ,$$

and for n by 1 vectors,

$$(A \Delta B)(\overline{C \Delta D}) = (A \overline{C}) \Delta (B \overline{D}) = (A \overline{D}) \Delta (B \overline{C}) \ , \tag{8}$$

$$(A \overline{B})^{\Delta M} = A^{\Delta M} \overline{B}^{\Delta M} \ ,$$

and, by definition, it is assumed that

$$(AD) \Delta (BC) = AD \Delta BC \ , \tag{9}$$

In the space of dimension 1, Δ is the regular complex scalar multiplication, i.e.,

$$t^{\Delta M} = t^{M} \ , \tag{10}$$

Similarly, the Δ exponential of a matrix A can be defined by:

$$e\Delta(A) = \sum_{M=0}^{\infty} A^{\Delta M}/M! \ , \tag{11}$$

where the Δ product has replaced the regular matrix product.

2.3 Mth-Order Spectral-Density Generating Function

Let an Mth-order spectral-density generating function be given by

$$\phi_{\Delta R1}(t) = E[e\Delta(R_1 \overline{R}_1(t))] \ . \tag{12}$$

Then, using definition (11) and properties (8)

$$\phi_{\Delta R1}(t) = \sum_{M=0}^{\infty} E[(R_1\bar{R}_1)^{\Delta M}] t^M/M! , \qquad (13)$$

where

$$D_{RM} = E[(R_1\bar{R}_1)^{\Delta M}] . \qquad (14)$$

is immediately identified. Thus

$$\phi_{\Delta R1}(t) = \sum_{M=0}^{\infty} D_{RM} t^M/M! , \qquad (15)$$

and hence the name of Mth-order spectral-density generating function.

2.4 **Linear Observation Equation**

Let the vector R_1 represent a linear measurement of a signal vector $S_1 = [S_{1,1}, \ldots, S_{i,1}, \ldots]$ in the following form:

$$R_1 = A_1S_1 + N_1 , \qquad (16)$$

where $n_1(t_k)$ is a vector of i.i.d noises, independent of the signal sequences $s_{i,1}(t_k)$. The signals $s_{i,1}$ are independent.

Using the independence of the signals with respect to the noises, the first-order spectral-density matrix D_{R1} is given by

$$D_{R1}(\omega) = A_1(\omega)E[S(\omega)\bar{S}(\omega)]\bar{A}_1(\omega) + E[N_1(\omega)\bar{N}_1(\omega)] , \qquad (17)$$

where

$$D_{S1} = E[S_1(\omega)\bar{S}_1(\omega)] , \qquad (18)$$

and

$$D_{N1} = E[N_1(\omega)\bar{N}_1(\omega)] , \qquad (19)$$

are readily identified; thus

$$D_{R1} = A_1 D_{S1} \bar{A}_1 + D_{N1} . \tag{20}$$

Building second-order signals, D_{R2} is computed as

$$D_{R2} = E[(R_1 \bar{R}_1)^{\Delta 2}] = E[(A_1 S_1 \bar{S}_1 \bar{A}_1 + A_1 S_1 \bar{N}_1 + N_1 \bar{S}_1 \bar{A}_1 + N_1 \bar{N}_1)^{\Delta 2}] . \tag{21}$$

Using the independence of the noises with respect to the sources, and for zero-mean sources and noises, terms as $E[A_1 S_1 \bar{S}_1 \bar{A}_1 \Delta A_1 S_1 \bar{N}_1]$ etc. average out. Moreover, terms as $E[N\Delta N]$ and $E[\bar{N}\Delta \bar{N}]$ are neglected compared to terms as $D_{N1} = E[N\Delta \bar{N}]$ since

$$E[n_{i,1} \Delta n_{i,1}] = \mathcal{F}(E[n_{i,1}(t) * n_{i,1}(t)]) , \tag{22}$$

where $\mathcal{F}$ is the Fourier transform operator, $E[n_{i,1}(t) * n_{i,1}(t)]$ is the correlation estimate between $n_{i,1}(t)$ and $n_{i,1}(-t)$, and for any time shift only one data point correlates.

Then D_{R2} equals

$$D_{R2} = E[(A_1 S_1 \bar{S}_1 \bar{A}_1)^{\Delta 2}] + 4E[A_1 S_1 \bar{S}_1 \bar{A}_1] \Delta E[N_1 \bar{N}_1] + E[(N_1 \bar{N}_1)^{\Delta 2}] \tag{23}$$

or

$$D_{R2} = E(A_1 S_1 \bar{S}_1 \bar{A}_1)^{\Delta 2}] + 4E[A_1 S_1 \bar{S}_1 \bar{A}_1] \Delta D_{N2} + D_{N2} . \tag{24}$$

Developing one element of the $\Delta 2$ product,

$$E[(A_1 S_1 \bar{S}_1 \bar{A}_1)^{\Delta 2}]_{ij} = E\left[\left(\left(\sum_{k1} a_{ik1} s_{k1} \right) \left(\sum_{k2} \bar{a}_{jk2} \bar{s}_{k2} \right) \right)^2 \right] \tag{25}$$

where a_{ij} is the element of A_1. Developing the square:

$$(25) = E\left[\sum_{k1} \sum_{k2} \sum_{k3} \sum_{k4} a_{ik1} a_{ik3} s_{k1} s_{k3} \bar{s}_{k2} \bar{s}_{k4} \bar{a}_{jk2} \bar{a}_{jk4} \right] . \tag{26}$$

For independent sources,

$$E[s_{k1} s_{k3} \bar{s}_{k2} \bar{s}_{k4}] = 0 \tag{27}$$

except in the following three cases:

i) $k1 \rightarrow k3$, $k2 \rightarrow k4$. Then it can be seen that:

$$(26) \rightarrow E\left[\sum_{k1}\sum_{k2} a_{ik1}^2 s_{k1}^2 \bar{s}_{k2}^2 \bar{a}_{jk2}^2\right] \rightarrow [A_2 D_{S2} \bar{A}_2]_{ij} , \tag{28}$$

where $A_2 \rightarrow A_1^{\Delta 2}$.

ii) $k1 \rightarrow k2 \neq k3 \rightarrow k4$. Then

$$(26) \rightarrow E\left[\sum_{k1}\sum_{k3} (a_{ik1} s_{k1} \bar{s}_{k1} \bar{a}_{jk1})(a_{ik3} s_{k3} \bar{s}_{k3} \bar{a}_{jk3})\right] \tag{29}$$

and since $k1 \neq k3$,

$$(26) \rightarrow E\left[\sum_{k1} a_{ik1} s_{k1} \bar{s}_{k1} \bar{a}_{jk1}\right] E\left[\sum_{k3 \neq k1} A_{ik3} s_{k3} \bar{s}_{k3} \bar{a}_{jk3}\right] \tag{30}$$

$$(26) \rightarrow [(A_1 P_1 \bar{A}_1)(A_1 P_1 \bar{A}_1)]_{ij} - E\left[\sum_{k3} a_{ik3}^2 s_{k3}^2 \bar{s}_{k3}^2 \bar{a}_{jk3}^2\right] . \tag{31}$$

Since the sources are independent, $[D_{SM}]_{ij} \rightarrow 0$ if $i \neq j$, and then,

$$(26) \rightarrow [(A_1 D_{S1} \bar{A}_1)(A_1 D_{S1} \bar{A}_1) - A_2 D_{S2} \bar{A}_2]_{ij} . \tag{32}$$

iii) $k1 \rightarrow k4 \neq k3 \rightarrow k2$; similar to ii).

Thus, adding the above three cases:

$$D_{R2} \rightarrow A_2 D_{S2} \bar{A}_2 + 2[(A_1 D_{S1} \bar{A}_1)^{\Delta 2} - A_2 D_{S2} \bar{A}_2] + 4(A_1 D_{S1} \bar{A}_1)\Delta D_{N1} + D_{N2} . \tag{33}$$

It can be easily verified that, in theory $D_{R2} \rightarrow D_{R1}^{\Delta 2}$ and $D_{S2} \rightarrow D_{S1}^{\Delta 2}$. Then, using (20), (33) simplifies into

$$D_{R2} \rightarrow A_2 D_{S2} \bar{A}_2 + 2[(D_{R1} - D_{N1})^{\Delta 2} - A_2 D_{S2} \bar{A}_2] + 4(D_{R1} - D_{N1})\Delta D_{N1} + D_{N2} \tag{34}$$

or

$$D_{R2} \rightarrow A_2 D_{S2} \bar{A}_2 + 2D_{N1}^{\Delta 2} - D_{N2} . \tag{35}$$

For independent, identically distributed Gaussian noises of coefficient β^2,

$$D_{N2} \rightarrow E[(N\bar{N})^{\Delta 2}] \rightarrow 4\beta^4 I , \tag{36}$$

and it yields

$$D_{R2} = A_2 D_{S2} \bar{A}_2 - \beta^4 I. \tag{37}$$

2.5 Mth-Order Eigenstructure Generating Function

More generally, for noise sequences with odd moments that are zero, such as the Gaussian noises, it can be shown that

$$D_{RM} = \sum_{p=0}^{M} A_p D_{Sp} \bar{A}_p \begin{pmatrix} P \\ M \end{pmatrix}^2 \Delta D_{N(M-p)} , \tag{38}$$

where $A_o D_{So} \bar{A}_o$ is the neutral matrix for the Δ-product and $\begin{pmatrix} P \\ M \end{pmatrix}$ is the binomial coefficient.

Let E_M be the Mth-order eigenstructure matrix defined as

$$E_M = A_M D_{SM} \bar{A}_M , \tag{39}$$

where E_M has the desired structure to apply eigenstructure type of methods, hence its name.

Then E_M can be computed from D_{RM} by a recurrent relation based on the two equations

$$E_1 = D_{R1} - D_{N1} ,$$

$$E_M = D_{RM} - \sum_{p=0}^{M-1} A_p D_{Sp} \bar{A}_p \begin{pmatrix} P \\ M \end{pmatrix} \Delta D_{N(M-p)} \tag{40}$$

For independent sequences, D_{NM} is a diagonal matrix, and thus E_M and D_{RM} differ only by their diagonal components.

For Gaussian i.i.d. noises,

$$D_{NM} = \frac{(2M)!}{M! \ 2^M} \beta^{2M} I , \tag{41}$$

and β^2 is estimated as the smallest eigenvalue of D_{R1}.

Then the eigenstructure generating function is conveniently defined as

$$\phi_{\square R1}(t) = \sum_{M=0}^{\infty} E_M t^M / M! = E[e\square(R_1 \bar{R}_1 t)] , \tag{42}$$

or

$$E_M - E[R_1\bar{R}_1)^{\square M} \tag{43}$$

where the $\square$ operator is defined a posteriori from relations (40) and (43).

3. APPLICATION TO THE DIRECTION-FINDING PROBLEM

3.1 The First-Order Eigenstructure Method

High resolution methods to solve the direction-finding problem have become increasingly popular since the early works of Schmidt [1] and Pisarenko [2] on the harmonic-retrieval problem. Successive improvements include the study of estimators for the location of the sources (Bienvenu and Kopp [3], Henderson [4]) and detection tests for the number of sources using likelihood methods (Rissanen [5], Wax and Kailath [6]).

n sources $s_{k,1}$, centered at frequency $\omega/2\pi$, are impinging on a linear array from directions ϕ_k. The linear array consists of d sensors located at distances D_i of sensor 1, such that $0 - D_1 < \ldots < D_d$. Then the signals received at sensor $r_{i,1}$ are

$$R_{i,1}(t) - \sum_{k=1}^{n} s_{k,1}(t-(D_i\sin\phi_k)/c) + n_{i,1}(t) , \tag{44}$$

where c is the speed of propagation, and the additive noises at the ith sensor, $n_{i,1}(t)$ are independent, identically distributed noises. Gaussianity is not required but will be used as an example. Techniques described by Paulraj and Kailath [7] can be applied when the noise field is unknown.

The Fourier transform of (44) yields

$$R_{i,1}(\omega) - \sum_{k=1}^{n} e^{-j\omega\mu_{ik}} S_{k,1}(\omega) + N_{i,1}(\omega) , \tag{45}$$

where $\mu_{ik} - D_i(\sin\phi_k)/c$. More generally, in matrix form,

$$R_1(\omega) - A_1(\omega)S_1(\omega) + N_1(\omega) , \tag{46}$$

and matrix $A_1(\omega)$ is the (first-order) direction matrix

$$
A_1(\omega) = \begin{bmatrix} 1 & . & . & . & 1 \\ e^{-j\omega\mu_{11}} & . & . & . & e^{-j\omega\mu_{1n}} \\ . & . & . & . & . \\ . & . & . & . & . \\ . & . & . & . & . \\ e^{-j\omega\mu_{d1}} & . & . & . & e^{-j\omega\mu_{dn}} \end{bmatrix} , \qquad (47)
$$

where the columns are the direction vectors of sources $s_{k,1}$:

$$
\vec{d}_k^T(\omega) = [1, e^{-j\omega\mu_{1k}}, \ldots, e^{-j\omega\mu_{dk}}] . \qquad (48)
$$

This yields the spectral-density generating function, which is herein referred to as the first order, according to section 2.4

$$
D_{R1} = A_1 D_{S1} \bar{A}_1 + \beta^2 I , \qquad (49)
$$

where β^2 is the noise spectral density coefficient of the i.i.d. noises.

Equation (49) has the desired structure to apply the eigenstructure method [2], [3]. If D_{S1} has rank n, i.e., if no two sources are coherent, the d by d spectral density matrix D_{R1} has n eigenvalues larger than β^2. The remaining d-n eigenvalues are equal to β^2 and have eigenvectors that are orthogonal to the columns of the direction matrix A_1, i.e., to the direction vectors. The same property can be used over the correlation matrixes. Then a simple orthogonality test allows computation of the receiving angles.

The direction-finding methods based on the eigenstructure rely on the hypothesis that the signals are not coherent, i.e., not fully correlated. Unfortunately, coherent sources occur frequently in practical problems, in case of jamming or in case of multipath propagation.

If two sources are coherent, e.g., $s_{2,1} = as_{1,1}$, then D_{S1} has rank n-1 and the d-n+1 eigenvectors are not orthogonal to the direction vectors as defined in equation (48) but are orthogonal to the d by n-1 direction matrix

$$
A_1 = \begin{bmatrix} 1+a & 1 & . & . & 1 \\ e^{-j\omega\mu_{11}} + ae^{-j\omega\mu_{12}} & e^{-j\omega\mu_{13}} & . & . & e^{-j\omega\mu_{1n}} \\ . & . & . & . & . \\ . & . & . & . & . \\ . & . & . & . & . \\ e^{-j\omega\mu_{d1}} + ae^{-j\omega\mu_{d2}} & e^{-j\omega\mu_{d3}} & . & . & e^{-j\omega\mu_{dn}} \end{bmatrix} \qquad (50)
$$

with a compound vector in the first column, as a function of three unknowns, namely ϕ_1, ϕ_2, and a. A_1 has one less column compared to (47).

If there are only d - n+1 sensors, then there are only two eigenvalues equal to β^2 with eigenvectors that are orthogonal to the compound direction vector. This provides two equations that cannot fix the three unknowns and the problem cannot be solved by this method. This results in an inconsistency in the eigenstructure method; though u sources are detected, the associated direction vectors are not proper. Then, for example, linear processing techniques performed by Henderson [4] propagate the singularity of the source spectral density matrix, and the methods described fail.

Shan, Wax, and Kailath [8] propose a method to recover from coherent sources by averaging spectral densities computed from different linear sub-arrays, taking advantage of the regular spacing of the sensors. This spacial smoothing approach trades off, in fact, half the effective aperture to recover from the possible coherence of the sources, resulting in a loss in both resolution and detection threshold.

The nonlinear approach developed above solves the case where all the sources that are not independent are coherent. It can be applied to any type of array and shows drastic improvements in terms of minimal aperture when the number of coherent sources is high. The nonlinear method also provides the coherence coefficients between the sources unlike the method by Shan, Wax, and Kailath [6].

3.3 The Second-Order Method

From section 2.4, the second-order eigenstructure matrix, i.e., the spectral-density matrix of the second-order signals, is readily given by (33), or, in the Gaussian case, by (37). These matrices have the required structure to apply the eigenstructure method, with A_2 as a direction matrix. The vector to test for orthogonality is

$$\vec{d}_2^T(\phi) = [1, e^{-2j\omega(D_2\sin\phi)/c}, \ldots, e^{-2j\omega(D_n\sin\phi)/c}] , \tag{51}$$

i.e., where each component in (47) (or in (50) when coherence is present) is squared.

Compared with the first order, the second order yields a second set of independent orthogonality conditions, function of the very same unknowns. Similarly, the third order, fourth, etc., would yield other sets of equations to solve the problem.

4. CONCLUSION

The Mth-order signals allow an enhancement of a coherent part of a signal with respect to the background noise. Though it is a pre-processing of information, any classical method must be adapted to the Mth-order signals in order to recover the information of the original signal. The direction finding example shows how straightforward some applications can be.

The major requirements are that the noises must be i.i.d. and independent from the signals. The second order is readily available to any signal processing technique with a linear observation equation. On the other hand, to be able to use higher-order eigenstructures, though Gaussianity is not enforced, powerful simplifications occur if the odd moments of the noises vanish.

ACKNOWLEDGEMENT

This research was supported by ONR Contract No. N00014-8L-K-0814 Mod. #P00005.

REFERENCES

1. R.O. Schmidt, "Multiple Emitter Location and Signal Parameter Estimation," *Proc. RDAC Spectral Estimation Workshop*, 243-258, 1979.
2. V.F. Pisarenko, "The Retrieval of Harmonics from a Covariance Function," *Geophys. J. Roy. Astron. Soc.*, Vol. 33, 247-266, 1973.
3. G. Bienvenu and L. Kopp, "Optimality of High Resolution Array Processing Using the Eigensystem Approach," *IEEE Trans. Acoustics, Speech and Signal Processing*, Vol. ASSP-31, 1235-1247, 1983.
4. T.L. Henderson, "Rank Reduction for Broadband Waves Incident on a Linear Receiving Aperture," *Proc. 19th Asilomar Conf. on Circ., Syst., & Computer*, Pacific Grove, CA, 1985.
5. J. Rissanen, "A Universal Prior for the Integers and Estimation by Minimum Description Length," *Ann. of Stat.*, Vol. 11, 417-431.
6. M. Wax and T. Kailath, "Detection of Signals by Information Theoretical Criteria," *IEEE Trans. Acoustics, Speech and Signal Processing*, Vol. ASSP-33, 387-392, 1985.
7. A. Paulraj, T. Kailath, "Eigenstructures Methods for Direction of Arrival Estimation in the Presence of Unknown Noise Fields," *IEEE Trans. Acoustics, Speech and Signal Processing*, Vol. ASSP-34, 13-20, 1986.
8. T.J. Shan, M. Wax, and T. Kailath, "Spatial Smoothing Approach for Location Estimation of Coherent Sources," *Proc. 18th Asilomar Conf. on Circ., Syst., & Computer*, Pacific Grove, CA, 1984.

STOCHASTIC CIRCULATORY LYMPHOCYTE MODELS

Z.H. Farooqi and R.R. Mohler
Dept. of Electrical and Computer Engineering
Oregon State University, Corvallis, OR 97331 (USA)

1. INTRODUCTION

Interest in mathematical immunology has been growing since its beginning as is reflected in the several monographs that have been published and the conferences that have been held lately. Examples of the work that has been done are available in Bell, Perelson, and Pimbley [1], Merrill [2], Mohler, Bruni, and Gandolfi [3], DeLisi [4], Marchuk [5], Marchuk and Belykh [6], Marchuk, Belykh, and Zuev [7], and Mohler [8]. The mathematical study of events that are involved at the cellular level in transmission of information seems to be generally missing in the litera-ture, the reference here being to the study of recirculation of lymphocytes. These events functionally interconnect many of the parts of the immune response by physically and biochemically conveying information and providing defense. This paper is concerned with the distribution of recirculating lymphocytes. Besides the healthy state of an organism such distribution is also of significance for disease states. Any maldistribution could be a symptom of a disease, for example, in humans, Hodgkin's disease, hepatitis B, psoriasis, and chronic lymphocytic leukemia. Thus quantification of the norm of distribution pattern, statistical analysis of the deviation from the norm and/or maldistribution, and modeling and identification of the models can be useful in diagnosis and estimation of disease severity, and in the classification and differential diagnosis of abnormalities in migratory patterns of lymphocytes.

In this paper, after a very brief review of deterministic models and discussion of the need of the presence of stochasticity, two stochastic models (a discrete-time and a continuous-time) are analyzed.

2. DETERMINISTIC MODELS

For details on the physiology of lymphocyte recirculation see Sprent [9] and DeSousa [10,11]. The experimental data used here was collected in the laboratory of and provided by the late Prof. W.L. Ford at the University of Manchester [12]. The basic network of organs used for the deterministic models is given in Figure 1.

Four deterministic models have been published, namely, linear time-invariant, linear time-variant, linear time-invariant time-delay, and nonlinear [13,14]. The

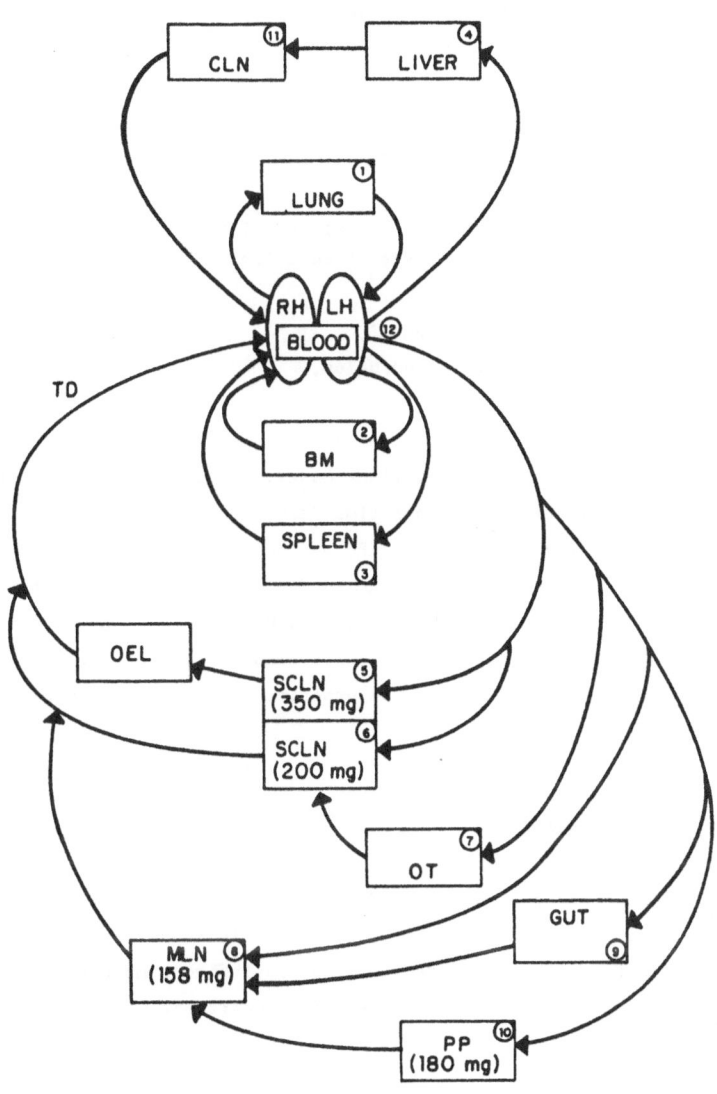

Figure 1. Lymphocyte Circulation.

CLN	:	Coeliac LN	SCLN:	Subcutaneous LN
TD	:	Thoracic Duct	OT :	Other Tissue
BM	:	Bone Marrow	MLN :	Mesenteric LN
OEL	:	Other Efferent Lymphatics	PP :	Peyers Patches
RH,LH:		Right, Left Heart		

general equation for the four models can be written in vector form as a retarded functional differential equation,

$$\dot{\underline{x}}(t) = A(t,\underline{x}(t))\ \underline{x}(t) + \sum_{i=1}^{m} B_i\ \underline{x}(t-\tau_i) \qquad (1)$$

where

> $\underline{x}(t)$ = state vector with components $x_i(t)$, $x_i(t)$ being the state of compartment i, which is the percentage of activity (labeled cells) relative to the injected dose,
>
> τ_i = time delay in compartment i,
>
> A, B_i = m×m matrices as explained below,
>
> m = number of compartments.

For the linear time-invariant model m = 12, $A(t,x(t))$ = A constant, and B_i = 0, τ_i = 0, $\forall i$. For the linear time-variant model m = 12, $A(t,x(t))$ = A(t), and B_i = 0, $\forall i$. In the linear time-invariant time-delay model m = 12, $A(t,x(t))$ = A constant, and B_1 = 0, $B_i \neq 0$, i = 2,...,12; while in the nonlinear case m = 9, $A(t,x(t))$ is a matrix function of x and t, and B_i = 0, $\forall i$. For details of the structure of $A(t,x(t))$ and B_i see Farooqi and Mohler [15]. Nonnegativity, boundedness, and the stability of these models are also discussed there. For a comparison of the states of these models with the experimental data see Mohler, Farooqi, and Heilig [14].

3. DESCRIPTION OF PARAMETERS

In [13,14] the parameters in the models (the elements of matrices A(.,.), B were simply explained as directional permeabilities. To reduce the uncertainty in the interpretation of parameters, first the factors influencing them will be discussed. Here, consider the flow of cells in a single compartment which satisfies Fick's principle [16]:

$$\frac{dm}{dt} = J_{in} - J_{out} , \qquad (2)$$

where

> m = mass,
>
> t = time,
>
> J_{in} = mass influx rate, and
>
> J_{out} = mass efflux rate.

The most important factors that influence lymphocyte migration are i) hemodynamic and ii) physico-chemical interaction between cells and vascular endothelium [17].

3.1 Hemodynamics

The whole blood behaves as a non-Newtonian fluid with the blood viscosity varying with the hematocrit (the percentage of blood volume occupied by cells). Blood flow in large vessels (i.e., those with diameter much larger than cell diameter) may be considered as homogeneous and obeying Hagen-Poiseuille's law; while microcirculation is characterized by low Reynold's number and nonhomogeneous flow with two phases (one plasma and the other cells). In the capillaries there is a continuous variation in blood flow velocity and the hematocrit is unsteady [18].

3.2 Leukocyte-Endothelial Interaction

While the erythrocytes tend to flow along the tubal axis, the leukocytes (including lymphocytes) prefer to stick to the vascular endothelium and roll along it [10]. This causes a shear, S, on the leukocyte which written as a dimensionless ratio, the shear coefficient, C_s, can be expressed in terms of other dimensionless constants [18]

$$C_S = f\left(\frac{d_c}{d_t}, \frac{V_c}{V_m}, N_R, H\right) , \tag{3}$$

where

d_c — diameter of WBC considered as a sphere,

d_t — diameter of blood vessel,

V_m — maximum velocity of undisturbed flow in the vessel,

V_c — linear velocity of centroid of the WBC

H — hematocrit,

$N_R = \rho V_m d_c/\mu$ — Reynolds number,

ρ — plasma density, and

μ — plasma coefficient of viscosity.

C_s is strongly dependent on H. Since H is stochastic in vivo, the shear fluctuates in time.

A leukocyte adhering to the endothelium is subjected to shear due to blood flow and also to an effective frictional drag. Thus accelerations and cell concentrations (c) are functions of distance along the vessel axis (z), thus there results a concentration gradient. By Fick's first law of diffusion a current, J, exists

$$J = -DA\partial C/\partial z \qquad \text{or} \qquad J = \alpha\Delta C , \tag{4}$$

where

    D = diffusion constant,

    A = cross-sectional area of the vessel.

Thus the coefficient α in (4) depends on several factors, some of which are stochastic in nature,

$$\alpha = f(D, A, S, F_{drag}, z) . \tag{5}$$

The J_{in} and J_{out} in (2) may thus be written in terms of cell concentrations, the coefficients being interpreted as functions of the various factors mentioned in (3) and (5).

4. THE DISCRETE MODEL

4.1 Development of the Model

For the purposes of constructing the stochastic models, the original network of interconnections (Figure 1) was simplified to a system with only 7 compartments, the connectivity diagram being shown in Figure 2.

Blood, lungs, spleen, and bone marrow were kept as separate compartments. All the others were lumped into three compartments, assuming homogeneity of structure and function within a compartment: i) tissues like gut, liver, peyer's patches, and miscellaneous nonlymphoid tissue as "other tissue," ii) the lymph nodes (LN) draining the other tissues (SCLN, MLN, CLN, etc.) as "LN-a," and iii) all the other LN as "LN-b."

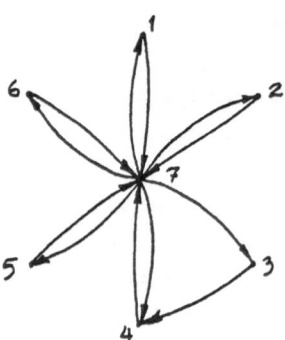

Figure 2. Connectivity diagram for the 7 compartmental model. (Key: 1 = bone marrow, 2 = spleen, 3 = "other tissue," 4 = LN-a, 5 = LN-b, 6 = lungs, 7 = blood.)

The discussion in section 3 was about blood vessels. Organs or compartments may be though of as vessels which sometimes behave as large ones and sometimes as microvessels (as in trapping). Thus each compartment possesses an effective diameter and an effective viscosity which are functions of its vasculature and the physiological microenvironment prevailing at that particular time. Thus for any compartment Eqs. (2) and (4) should be satisfied. Because of the presence of inherent stochasticity and the lack of precise knowledge of various variables, the model equations (from cell concentration) in discrete-time written in vector form comprise a bilinear time series:

$$X_k = A \; X_{k-1} + \beta_{k-1} \; X_{k-1} + \mathcal{E}_k , \tag{6}$$

where

$$X_k = [x_{1k}, \ldots, x_{7k}]'$$
$$\mathcal{E}_k = [\epsilon_{1k}, \ldots, \epsilon_{7k}]'$$

' stands for transposition.

$$A = \begin{bmatrix} 1-\alpha_1 & & & & & & a_{17} \\ & 1-\alpha_2 & & & & & a_{27} \\ & & 1-\alpha_3 & \phi & & & a_{37} \\ & & \alpha_3 & 1-\alpha_4 & & & a_{47} \\ & \phi & & & 1-\alpha_5 & & a_{57} \\ & & & & & 1-\alpha_6 & a_{67} \\ \alpha_1 & \alpha_2 & 0 & \alpha_4 & \alpha_5 & \alpha_6 & 1-\alpha_7 \end{bmatrix} , \tag{7}$$

$$\alpha_7 = \sum_{i=1}^{6} a_{i7} , \tag{8}$$

$$\beta(n) = \begin{bmatrix} -\beta_1(n) & & & & & & b_{17}(n) \\ & -\beta_2(n) & & & & & b_{27}(n) \\ & & -\beta_3(n) & \phi & & & b_{37}(n) \\ & & \beta_3(n) & -\beta_4(n) & & & b_{47}(n) \\ & \phi & & & -\beta_5(n) & & b_{57}(n) \\ & & & & & -\beta_6(n) & b_{67}(n) \\ \beta_1(n) & \beta_2(n) & 0 & \beta_4(n) & \beta_5(n) & \beta_6(n) & -\beta_7(n) \end{bmatrix} , \tag{9}$$

$$b_{17}(n) = \frac{a_{17}}{\alpha_7} \beta_7(n) \text{ such that } \beta_7(n) = \sum_{i=1}^{6} b_{i7}(n) , \tag{10}$$

$$\epsilon_7(n+1) = \sum_{i=1}^{6} \epsilon_i(n+1) . \tag{11}$$

$x_i(n)$ — state of compartment i at time instant n, being the number of lymphocytes in compartment i expressed as a percent of the total number in
the system as measured by the presence of radioactive label.

α_i — deterministic part of transfer parameters (leaving i), a_{i7} is portion of
α going from 7 to i. α_i (and a_{i7}) scales the percent activity in
compartment i (or 7) at instant n (dimensionless).

$\beta(n)$ — stochastic part of transfer parameters (leaving i), $b_{i7}(n)$ is the
portion of $\beta_7(n)$ going from 7 to i. $\{\beta_i(n)\}$ are i.i.d. and constitute
multiplicative noise (dimensionless).

$\epsilon(n)$ — additive noise, added to the input and/or output of the compartment;
$\{\epsilon_i(n)\}$ are i.i.d. (dimensions of percent activity).

$\{\beta_i(n)\}$ and $\{\epsilon_i(n)\}$ are independent.

The subscripts are as follows:

1 — bone marrow,   2 — spleen,
3 — other tissue,   4 — LN-a,
4 — LN-b,   6 — lungs,
7 — blood.

It is assumed that there are no births and deaths, and no loss of label.
Experimentally, the number of cells in any compartment is measured by the amount of
radiolabel activity. The 7×7 matrix A is constant. The 7×7 matrix β_k is compartmental and is interpreted as the random component in (5). $\underline{\epsilon}_k$ is a random 7-
vector. Presence of $\underline{\epsilon}_k$ in (6) may be justified by changes in the same variables as
for β_k, but these are due to intercompartmental stochasticity rather than intra-
compartmental randomness (as for $\beta(n)$). Also $\underline{\epsilon}(n)$ may mimic the presence of
different types of immunities and other physiological factors, and the random
variation in different organisms within the same species.

Equation (6) can be viewed as a system of difference equations with both slope
and intercept random or as a Markov chain generated by two stochastic processes
(i.e., a Markov chain in a random environment), but such processes will not be
discussed here. The second term on the right-hand side of (6) causes the structure
to be similar to that of the bilinear time series studied by Tang and Mohler in this
volume.

Considerable literature is available on bilinear time series with one and two
inputs [19,20]. A bilinear time series with two inputs has sometimes been called
random coefficient autoregression. It is generally assumed that, in (6), A is a
constant matrix, β_k is a matrix such that $E[\beta_k] = 0$ and $E[\beta_k \otimes \beta_k] = C_\beta$, and $\underline{\epsilon}_k$ is
such that $E[\underline{\epsilon}_k] = \underline{0}$ and $E[\underline{\epsilon}_k \underline{\epsilon}_k] = G$. $\{\beta_{ik}\}$ and $\{\epsilon_{ik}\}$ are independent. $\otimes$, the
Kronecker product, is defined such that the ij-th block of A⊗B is $[A \otimes B]_{ij} = a_{ij}B$.
Equation (6) may be used to find the mean and variance of the state. Results on
second-order stability, second-order stationarity, their relationship, strict

stationarity, and asymptotic properties of least squares and maximum likelihood
estimators are given in Nicholls and Quinn [19]. Their biological interpretation
that is relevant here follows.

If the variance in the number of recirculating lymphocytes in any compartment
reaches an equilibrium which does not depend on the number of cells present ini-
tially in the compartment, then the mean number of cells also reaches an equilibrium. In other words, at equilibrium, the distribution of cells in that compartment has attained a constant mean and standard deviation. When the random perturbations in the recirculating lymphocyte pool are independent and identically distributed, the necessary and sufficient condition for the lymphocyte distribution in
any compartment to have a constant mean and standard deviation at equilibrium is
that the variation in the number of lymphocytes there be stable. Under these
conditions the distribution is also asymptotically time-invariant. This implies
that the number of recirculating lymphocytes can be characterized in terms of their
population distributions in different compartments.

4.2 Estimation and Statistical Analysis

Estimates for the elements of A were obtained using the equation for means and
those for the variances of the random variables from

$$\underline{P}_{d_k} = A_d \underline{P}_{d_{k-1}} + C_d \underline{P}_{e_k} \tag{12}$$

where

$\underline{P}_{d_k}$ = a 7-vector, diagonal of the variance-covariance matrix,

$$A_d = \begin{bmatrix} (1-\alpha_1)^2+\beta_1^2 & & & & & & a_{17}^2+b_{17}^2 \\ & (1-\alpha_2)^2+\beta_2^2 & & & & & a_{27}^2+b_{27}^2 \\ & & (1-\alpha_3)^2+\beta_3^2 & & \phi & & a_{37}^2+b_{37}^2 \\ & & \alpha_3^2+\beta_3^2 & (1-\alpha_4)^2+\beta_4^2 & & & a_{47}^2+b_{47}^2 \\ & \phi & & & (1-\alpha_5)^2+\beta_5^2 & & a_{57}^2+b_{57}^2 \\ & & & & & (-\alpha_6)^2+\beta_6^2 & a_{67}^2+b_{67}^2 \\ \alpha_1^2+\beta_1^2 & \alpha_2^2+\beta_2^2 & 0 & \alpha_4^2+\beta_4^2 & \alpha_5^2+\beta_5^2 & \alpha_6^2+\beta_6^2 & (1-\alpha_7)^2+\beta_7^2 \end{bmatrix},$$

$$
C_d = \begin{bmatrix}
\sigma_{\epsilon 1} & & & & & & \\
& \sigma_{\epsilon 2} & & & & & \\
& & \sigma_{\epsilon 3} & & \phi & & \\
& & & \sigma_{\epsilon 4} & & & \\
& \phi & & & \sigma_{\epsilon 5} & & \\
& & & & & \sigma_{\epsilon 6} & \\
& & & & & & \sigma_{\epsilon 7}
\end{bmatrix} ,
$$

and

$$
\underline{P}_{\underline{e}_k} = \mathrm{diag}(E[\underline{\epsilon}_k \underline{\epsilon}'_k]) .
$$

Some important covariances, like those between blood and the other compartments, are not available in the experimental data. Equation (12) does not take the important covariances into account.

The weighted least squares cost function $Q_{\chi^2}(\cdot) = \frac{1}{nk} \sum_{i=1}^{n} \sum_{j=1}^{k} \frac{(x_{ij} - \hat{x}_{ij})^2}{\sigma_{ij}^2}$ was used. It is a function of the sample variance of the residual errors, so that effectively this is what is minimized to obtain the "best" parameter estimates. Being divided by variance from the data as weights, Q.(.) is a dimensionless ratio. For the first moments min $Q_{\chi^2}(\hat{A}) = 5.73$, while for the second moments min $Q_{\chi^2}(\hat{C}_\beta, \hat{G}) = 16428.22$. The estimated parameter values are tabulated in Table 1. The minimized χ^2-value is significant which indicates that the model residual errors are quite large.

4.2.1 Stability

First-Moments

A is an irreducible matrix. It is termed column stochastic, and its spectral radius is 1. This is because the seven compartments are not linearly independent since blood is the sum of the other compartments, and overall it is a closed system.

Table 1. Estimated Parameter Values for the Discrete Model
Weighted Least-Squares Estimation

Deterministic Parts of Multiplicative Parameters:

α_1	a_{17}	α_2	a_{27}	α_3
.11948E-01	.48594E-02	.78494E-02	.65323E-01	.15659E-02

a_{37}	α_4	a_{47}	α_5	a_{57}
.35198E-01	.33771E-02	.34741E-02	.41243E-03	.36640E-02

α_6	a_{67}	α_7
.38460E+00	.39131E+00	.49617E+00

Standard Deviations of the Multiplicative Noises:

$\sigma_{\beta 1}$	$\sigma_{\beta 2}$	$\sigma_{\beta 3}$	$\sigma_{\beta 4}$	$\sigma_{\beta 5}$
.10245E-02	.99294E+00	.99673E+00	.15762E-03	.10000E+01

$\sigma_{\beta 6}$	$\sigma_{\beta 7}$
.93364E+00	.29113E+0

Standard Deviations of the Additive Noises:

$\sigma_{\epsilon 1}$	$\sigma_{\epsilon 2}$	$\sigma_{\epsilon 3}$	$\sigma_{\epsilon 4}$	$\sigma_{\epsilon 5}$
.23895E-04	.19190E-05	.22762E+01	.56450E-05	.74743E-01

$\sigma_{\epsilon 6}$	$\sigma_{\epsilon 7}$
.37201E-06	.23510E+01

Second-Moments

The matrix (A⊗A + C_β) is a 49×49 sparse matrix and the usual method for finding eigenvalues are not efficient in such cases. Ranges of eigenvalues were estimated using Gershgorin's Circle Theorem [21] and the fact that the spectral radius of a matrix is always greater than or equal to the largest diagonal element. It turned out that the spectral radius was greater than 1. (The eigenvalues lie in the interval [-6.27,8.5] and the spectral radius > 1.992).

The implication is of second-order instability and second-order non-stationarity [19]. This being so and the minimized χ^2 being significant at α = 0.01, further analysis was done to check into the goodness of fit of the compartments of the model. Since one of the conditions for existence and convergence of second moments, $\rho(A\otimes A + C_\beta) < 1$, is the same as the condition for stability which is violated, the second moment diverges [22]. There is a possibility of this happening because the variances were estimated by using only the diagonal of the $E[\underline{x}_k\underline{x}_k']$ matrix, covariances not being available in the data. By doing so the number of equations used in the estimation of the parameters is reduced and in turn the number of constraints on the parameters is also reduced thus leading to 'misestimation,'

4.2.2 Statistical Analyses

Since here the exact distributions, F_i, are not known, the following null and alternative hypotheses were tested on the residual errors between the model output and the experimental data:

H_o: $F_1 = F_2 = \ldots = F_7$

H_1: Not all F_i are equal

Friedman's two-way analysis of variance based on ranks was done. The null hypothesis concerns distributions (not means). For the first moments the test statistic $\hat{\chi}_R^2$ = 25.68 was obtained, which is significant (χ^2 = 16.81, k = 7, α = 0.01). This suggests that the errors in all the compartments do not have the same distributions. Wilcoxon and Wilcox multiple comparison procedure was used to do all pairwise comparisons. It suggests that the distribution of errors in compartment 2 is significantly different from the distributions in compartments 1, 3, 4, 5, and 7, as seen from Table 2.

For the second moments also, Friedman's analysis showed differences in distributions ($\hat{\chi}_R^2$ = 44.60 > χ^2 = 16.81, k = 7, α = 0.01). Wilcoxon and Wilcox test suggests three subsets of compartments with similar distributions

Table 2.   Statistical Analyses Discrete Time Model.

First Moments, Weighted Least-Squares Estimation

I(a): Friedman Test

Compartment Number	Sum of Ranks
1	59
2	24
3	64
4	58
5	63
6	34
7	62

$$\hat{\chi}_R^2 = 25.68**$$

I(b): Wilcoxon and Wilcox Test

From tables: $D_{WW;\alpha} = \begin{array}{ll} 32.5 & \alpha = 0.05 \\ 38.0 & \alpha = 0.01 \end{array}$

	5 (63)	7 (62)	1 (59)	4 (58)	6 (34)	2 (24)
3 (64)	1	2	5	6	30	40**
5 (63)		1	4	5	29	39**
7 (62)			3	4	28	38**
1 (59)				1	25	35*
4 (58)					24	34*
6 (34)						10

Second Moments; Weighted Least-Squares Estimation

I(a): Friedman Test

Compartment Number	Sum of Ranks
1	50
2	72
3	82
4	17
5	58
6	44
7	41

$$\hat{\chi}_R^2 = 45.33**$$

I(b): Wilcoxon and Wilcox Test

From tables: $D_{WW;\alpha} = \begin{array}{ll} 32.5 & \alpha = 0.05 \\ 38.0 & \alpha = 0.01 \end{array}$

	2 (72)	5 (58)	1 (50)	6 (44)	7 (41)	4 (17)
3 (82)	10	24	32	38**	41**	65**
2 (72)		14	22	28	31	55**
5 (58)			8	14	17	41**
1 (50)				6	9	35*
6 (44)					3	27
7 (41)						24

*indicates significance at $\alpha = 0.05$
**indicates significance at $\alpha = 0.01$

{(1,2,3,5),(1,2,5,6,7),(4)}

It should be cautioned here, however, that the basic assumptions (for validity of these tests relative to the limited available experimental data) are violated. Still they should provide some indication.

4.2.3 Biological Interpretation of the Results

It was shown above that the estimated parameter values the model mean is marginally stable. Stability of the second moment is a function of the squares of the deterministic parts of the parameters and the variances of the parametric noise. Because of the parametric noise, the model is a variable structure system (actually structurally unstable system) with random variation in the structure. As the estimated variances of the noises have large values for the model, the instability in the distribution of recirculating lymphocytes is amplified. Comparing lymphocyte populations in any compartment there is a big difference in the estimated number of cells and the experimental quantity. For the mean, compartment 2 (spleen) appears to be the main source of error, while for the variance (standard deviation), compartments 3 (LT) and 4 (LN-a) may be the main cause.

5. THE CONTINUOUS-TIME MODEL

With the same physiological reasoning as in section 3 and following the mathematical reasoning of Zuev [7,23], the general structure obtained for the continuous-time model is

$$d\underline{x}(t) = A(\underline{\ell},\underline{x}(t,\underline{\ell})) \ \underline{x}(t) \ dt + G(\underline{\ell},\underline{x}(t,\underline{\ell})) \ dW(t) \ . \tag{14}$$

Specializing this for our purposes, a model homologous to the discrete-time one (ignoring additive noise) is

$$d\underline{x}(t) = A\underline{x}(t)dt + \sum_{i=1}^{m} B_i\underline{x}(t)dW^i(t) \ ,$$

$$\underline{x}(t_0) = c \ , \ t\epsilon[t_0,\tau] \ . \tag{15}$$

The existence, uniqueness, and asymptotic properties of the solution of (14) and (15) are available in the literature (e.g., Arnold [24]).

The ordinary differential equations satisfied by the first and second moments of $\underline{x}(t)$ in (15) were used in parameter estimation. Due to lack of space, details are not given here. Interpretation of the results obtained after estimation and analysis is quite similar to the case of the discrete model. It can be shown that for the mean values, the main source of error lies in compartments 2 and 3, while for the variance it is in compartments 1 and 3 [25].

6. CONCLUSION AND FUTURE RESEARCH

It can be shown that the deterministic parameters of the two models are related in the sense that with closer samples the discrete-time model converges to the continuous-time model. But the second moments in both models diverge for the estimated values of the parameters. Does this mean that the models are not valid?

Many assumptions were made in the construction of the models which are violated in reality: homogeneity of organs, independence of events, stationarity, no births and deaths, indistinguishable cells, well-mixed compartments, same probability of transition for cells, etc. As in most biological experiments, the data available here is sparse. Better estimation procedures need to be used. Here the attempt was to keep procedures as simple and straightforward as possible so that they are also easily interpretable by biologists.

Personally, the author believes that, in spite of the behavior exhibited in the models here, with some changes they can be made to mimic the real data closely. A death term for lymphocytes needs to be included and probably an additive noise should be used to account for random variations between organisms. The likely compartments which need more attention are bone marrow, spleen, and lumping together of organs into "other tissues" which should be divided into subcompartments which are histologically compatible. Perhaps the biggest problem, however, is the experiment itself. While a recirculating pool of cells must be conservative (except for a small number of deaths during the period in test), it is apparent from the data that the "other tissue" compartment must account for an increasing number of unaccountable cells (radioactivity) as time evolves. Maximum likelihood estimation and simulation using non-Gaussian (e.g., lognormal) distribution needs to be done.

The models and the experiment presented here are just a first step. A great deal of work needs to be done to derive models that would approximate the real world more closely. In spite of their shortcomings, the models presented here do reflect a glimpse of the real system, namely, the closeness of the immune system to structural instability. Actually, this makes the real system more adaptable (controllable) in its quest to control localized infection.

ACKNOWLEDGEMENT

This research was supported by NSF Grant ECS 8215724.

REFERENCES

1. G.I. Bell, A.S. Perelson, and G.H. Pimbley, Jr., "Theoretical Immunology,"
    (Immunology Series, Vol. 8), Marcel Dekker, New York, 1978.
2. S.J. Merrill, "Mathematical Models of Humoral Immune Response," in T.A. Burton,
    (Ed.), "Modeling and Differential Equations in Biology," (Lecture Notes in Pure
    and Applied Mathematics), Marcel Dekker, New York, 13-50, 1980.
3. R.R. Mohler, C. Bruni, and A. Gandolfi, "A Systems Approach to Immunology,"
    Proc. IEEE 68(8), 964-990, 1980.
4. C. DeLisi, "Mathematical Modeling in Immunology," Ann. Rev. Biophys. Bioeng. 12,
    117-138, 1983.
5. G.I. Marchuk, Mathematical Models in Immunology, (Translation Series in Mathe-
    matics and Engineering), Optimization Software, New York, 1983.
6. G.I. Marchuk and L.N. Belykh, (Eds.), Mathematical Modeling in Immunology and
    Medicine, Proc. IFIP-TC 7 Working Conf. on Math. Modeling in Immunology and
    Medicine, Moscow, USSR, North-Holland, Amsterdam, 1982.
7. G.I. Marchuk, L.N. Belykh, and S.M. Zuev, 1985, "Mathematical Modelling of
    Infectious Diseases," In G.I. Marchuk and V.P. Dymnikov, (Eds.), Problems of
    Computational Mathematics and Mathematical Modelling, Advances in Science and
    Technology in U.S.S.R. (Mathematics and Mechanics Series), Mir Publishers,
    Moscow, pp 223-240, 1985.
8. R.R. Mohler, "Foundations of Immune Control and Cancer," A.V. Balakrishnan,
    (Ed.), Recent Advances in System Science, Optimization Software, Inc., (to
    appear), 1987.
9. J. Sprent, "Recirculating Lymphocytes," In J.J. Marchalonis, (Ed.), The Lym-
    phocyte: Structure and Function, Part 1, (Immunology Series, Vol. 5), Marcel
    Dekker, New York, pp 43-111, 1977.
10. M. DeSousa, "Cell Traffic," In P. Cuatrecasas and M.F. Greaves, (Eds.), "Recep-
    tors and Recognition," Vol. 2, Series A, Chapman and Hall, London, pp. 103-163,
    1976.
11. M. DeSousa, "Lymphocyte Circulation: Experimental and Clinical Aspects,", John
    Wiley and Sons, New York, 1981.
12. M.E. Smith and W.L. Ford, "The Recirculating Lymphocyte Pool of the Rat: A
    Systematic Description of the Migratory Behavior of Circulating Lymphocytes,"
    Immunol. 49, pp. 83-94, 1983.

13. R.R. Mohler, Z. Farooqi, and T. Heilig, "An Immune Lymphocyte Circulation Model," In P. Thoft-Christensen, (Ed.), Systen Modelling and Optimization, Proc. 11th IFIP Conference, Copenhagen, Denmark, 1983, (Lecture Notes in Control and Information Sciences, Vol. 59), Springer-Verlag, Berlin, 697-702, 1984.

14. R.R. Mohler, Z. Farooqi, and T. Heilig, "Lymphocyte Distribution and Lymphatic Dynamics," In A.V. Balakrishnan, A.A. Dorodnitsyn, and J.L. Lions, (Eds.), 1986, Vistas in Applied Mathematics: Numerical Analysis, Atmospheric Sciences, Immunology, Optimization Software, Inc., Publications Division, New York, 317-333, 1986.

15. Z. Farooqi and R.R. Mohler, "Distribution of Recirculating Lymphocytes. I: A Review of the Deterministic Models," in preparation, 1987.

16. S.E. Rubinow, Introduction to Mathematical Biology, John Wiley and Sons, 1975.

17. W.L. Ford, M.E. Smith, and P. Andrews, "Possible Clues to the Mechanism Underlying the Selective Migration of Lymphocytes from the Blood," In A. Curtis, (Ed.), Cell-Cell Recognition, Symposia So. Exp. Bio. XXXII, Cambridge University Press, Cambridge, 359-392, 1978.

18. Y.C. Fung, Biodynamics: Circulation, Springer Verlag, New York, 1984.

19. D.F. Nicholls and B.G. Quinn, Random Coefficient Autoregressive Models: An Introduction, (Lecture Notes in Statistics, Vol. 11), Springer-Verlag, New York, 1982.

20. T. Subba Rao and M.M. Gabr, An Introduction to Bispectral Analysis and Bilinear Time Series Models, Lecture Notes in Statistics, Vol. 24, Springer-Verlag, New York, 1984.

21. P. Lancaster, "Theory of Matrices," Academic Press, New York, 1969.

22. P.D. Feigin and R.L. Tweedie, "Random Coefficient Autoregressive Processes: A Markov Chain Analysis of Stationarity and Finiteness of Moments," J. Time Series Anal. 6(1), 1-14, 1985.

23. G.I. Marchuk, A.L. Asachenkov, L.N. Belykh, and S.M. Zuev, "Mathematical Modelling of Infectious Disease," In G.W. Hoffman and T. Hraba, (Eds.), Immunology and Epidemiology, Proc. Int'l Conf., Mogilany, Poland, (Lecture Notes in Biomathematics, Vol. 65), Springer-Verlag, Berlin, 64-81, 1985.

24. L. Arnold, Stochastic Differential Equations: Theory and Applications, John Wiley, New York, 1974.

25. Z. Farooqi and R.R. Mohler, "Distribution of Recirculating Lymphocytes. II: The Stochastic Models," in preparation, 1987.

Lecture Notes in Control and Information Sciences

Edited by M. Thoma and A. Wyner

Lecture Notes in Control and Information Sciences

Edited by M. Thoma and A. Wyner

Lecture Notes in Control and Information Sciences

Edited by M. Thoma and A. Wyner